**Published by**

# Land Rover Ltd

**A Managing Agent for Land Rover UK Limited**

**PUBLICATION NO. LSM 0054 HB (EDITION 7BB)**
**Copyright Land Rover 1987 and 1992**

# Land Rover
# NINETY & ONE TEN
# Handbook

**Covering Model years 1983 to 1990 inc.**

# CONTENTS

## Models covered

The information in this handbook covers current versions of Land Rover 'Ninety' and 'One Ten' petrol and diesel models. It is presented in sections to guide the reader progressively from reception of the vehicle through familiarisation with controls and instrumentation, driving techniques, basic day to day attention and longer term workshop maintenance. The final section lists technical data, recommended lubricants and fluids and electrical data.

Advice on operating turbo-charged diesel models is also included in this book. On these models, it is particularly important that the proper engine starting and stopping procedures are followed and that the special engine lubricating oils are used.

Where specific information is sought, first consult the list of contents (at the front of the book) which will direct you to the relevant page or pages.

## Warnings

For the protection of yourself and others and the longer service life of your vehicle please heed the instructions carefully and note the **WARNINGS** and **CAUTIONS** that are given throughout this handbook in the following form:

 **WARNING: Procedures which must be followed precisely to avoid the possibility of personal injury.**

**CAUTION:** This calls attention to procedures which must be followed to avoid damage to components.

**NOTE:** This calls attention to methods which make a job easier to perform.

 **WARNING: DO NOT mix Cross-Ply and Radial-Ply tyres on this vehicle. Recommended tyre replacements are given on the inside rear cover of this book.**

## Safety

In the interests of road safety, your attention is drawn to the following important safety hints.

1. Regular servicing, including day to day attention by the driver/owner as described in Section 4 and the longer term workshop maintenance described in Section 5, is essential to help provide safe, dependable and economical motoring and to ensure that the vehicle conforms to the various safety regulations in force.
2. Always use the seat belts, even for the shortest journeys.
3. Before driving, learn the layout and use of all controls, gears and switches.
4. Adjust the seat as necessary to achieve a comfortable driving position with full control over the vehicle.
5. Always start vehicle and operate controls from the driving position.
6. Ensure that the vehicle speed is low enough for an emergency stop to be made safely under all road and vehicle loading conditions.
7. Keep the windscreen, rear and side windows clean to give clear vision. Use a solvent in the screen washer reservoir.
8. Maintain all external lights in good working order and correct setting of headlamp beams.
9. When a steering lock is fitted, DO NOT turn the ignition key to the lock position or try to remove the key whilst the vehicle is in motion.

10. When fitted, ensure the power take-off (pto) universal joints are shielded.
11. Before working on pto driven implements always disengage the pto and switch 'off' the engine.
12. Maintain correct tyre pressures. These should be checked at least each month, or more often when high-speed touring or under cross-country conditions, even to the extent of a daily check.

 **WARNING: DO NOT remove the expansion tank filler cap when the engine is hot, because the cooling system is pressurised and personal scalding could result.**

 **WARNING: Many liquids and other substances used in motor vehicles are poisonous, they must not be consumed under any circumstances and must be kept away from open wounds.**

**These substances include anti-freeze, brake fluid, fuel, windscreen washer additives, lubricants, battery contents and various adhesives.**

**Key numbers on models fitted with steering column lock**
For security reasons the key numbers are not marked on the locks. If the key for the steering column lock is lost, the vehicle cannot be driven. For this reason and because the keys are of a special type, obtainable only from a Land Rover Distributor or Dealer, two steering column lock keys are supplied with each vehicle.
Owners are advised to take the following action:
(a) Immediately on receipt of the vehicle, record all the key numbers so that in case of loss, new keys can be obtained.
(b) Keep a spare steering column lock key away from the vehicle in a safe place, but where it is readily accessible.
The steering column lock if properly used reduces the possibility of theft.

**Vehicle Identification Number (V.I.N.)**
The Vehicle Identification Number and the recommended maximum vehicle weights are stamped on a plate that is rivetted to the top of the brake pedal box in the engine compartment.
The number is also stamped on the right-hand side of the chassis forward of the spring mounting turret.
Always quote this number when writing to Land Rover Limited or your Distributor and Dealer on any matter concerning your Land Rover.

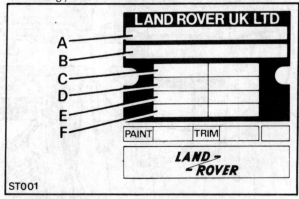

ST001

**Key to Vehicle Identification Number Plate – Fig. ST001**
A   Type approval
B   V.I.N. (minimum of 17 digits)
C   Maximum permitted laden weight for vehicle
D   Maximum vehicle and trailer weight
E   Maximum road weight — front axle
F   Maximum road weight — rear axle

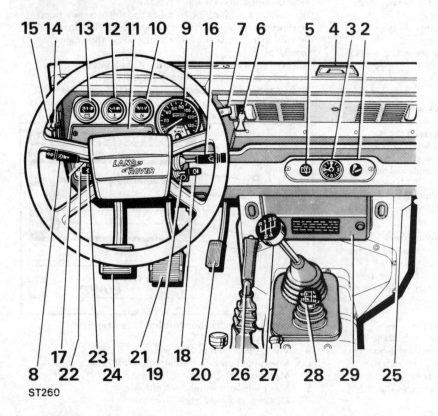

ST260

1 Ventilator control
2 Cigar lighter (option)
3 Clock (option)
4 Ash tray
5 Rear screen wash/wipe switch (option)
6 Ventilator control
7 Heater fan control
8 Headlamp dip, direction indicators, horn and flasher switch
9 Speedometer
10 Fuel gauge
11 Warning light cluster
12 Water temperature gauge
13 Voltmeter (option)
14 Heat temperature control
15 Heat distribution control
16 Windscreen washer and wiper switch
17 Switch panel for hazard warning, instrument and interior lighting and heated rear screen (option)
18 Rear fog guard lighting switch
19 Choke control (Petrol models)
20 Accelerator pedal
21 Footbrake pedal
22 Starter and steering lock switch
23 Main lighting switch
24 Clutch pedal
25 Bonnet release handle
26 Transmission handbrake lever
27 Main gearchange lever
28 Transfer gear/differential lock lever
29 Fuse box
30 Footwell vent

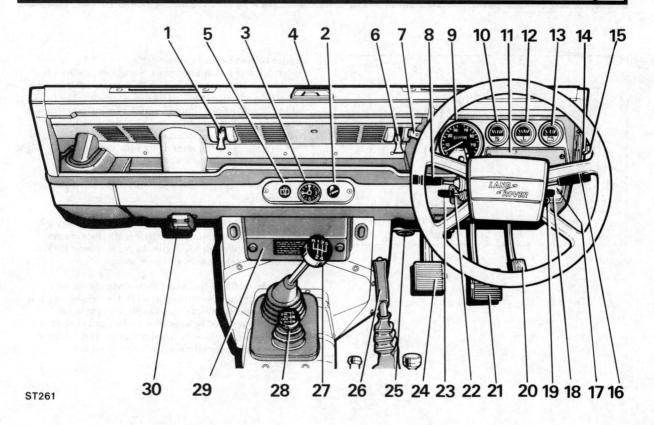

ST261

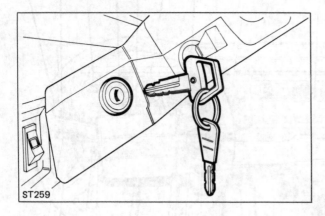

ST259

**Steering column lock (where applicable)**

On models fitted with a steering column lock, the lock is an integral part of the ignition switch on petrol models and the heater plug and starter switch on diesel vehicles. The following instructions should be studied in conjunction with the engine starting procedure below and overleaf.

To unlock the steering, insert the key and turn it forward to the first position. If the steering lock has been engaged, slight movement of the steering wheel will assist in its disengagement.

To lock the steering, turn the key fully back, and withdraw it from the lock.

⚠️ **WARNING: If for any reason the (ignition) engine is switched off while the vehicle is in motion, do not attempt under any circumstances to remove the key, otherwise the steering lock will be engaged.**

⚠️ **WARNING: To prevent the steering column lock engaging it is most important that before the vehicle is moved in any way, for example, being towed or coasting, the key must be inserted in the lock and turned to the first position. If, due to an accident or electrical fault it is not considered safe to turn the key, the battery must first be disconnected.**

### Ignition and starter switch (Petrol models)

1. The ignition switch has four positions:
   (A) Key upright; switch off.
   (B) With the ignition key turned to the first position, the heater blower motor and accessories, such as a radio, can be used.
   (C) Continue to turn the key to the second position to switch on the ignition.
   (D) Turn the key further against spring pressure to operate the starter motor. The key will automatically return to the second position when released.

In cold weather, depress the clutch pedal while the starter motor is in operation to improve engine starting speed.

### Cold start control

2. For ease of starting during winter and summer, the control should be fully pulled out and locked in position by turning the knob slightly. This action will progressively enrich the mixture for cold starting.

   After the engine has started, return the control to the off (fully pushed in) position, as soon as possible consistent with even running.

   A fast idle engine speed can be obtained by setting the control within the first 12 mm ($\frac{1}{2}$ in) of its initial movement. This position increases the engine speed without an alteration in mixture.

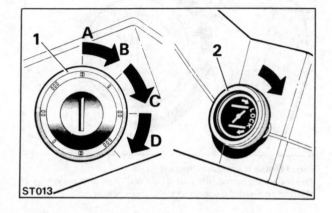

ST013

### Starting with a warm engine

DO NOT use the cold start control or pump the accelerator pedal. Depress the accelerator pedal to approximately half-way. Turn the ignition key to operate the starter, keeping the accelerator pedal in the half-way position. Release the ignition key and accelerator pedal immediately the engine starts.

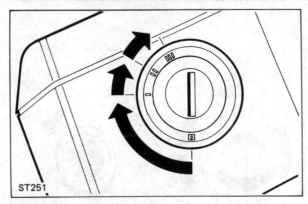

ST251

### Diesel engine starting and stopping
The following procedures must be used to ensure easy starting and avoid damage to the turbo-charger.

### Heater plugs
The Land Rover diesel engine will start satisfactorily, with the proper use of the heater plugs, down to temperatures of −20°C (−4°F) even with batteries only 70% charged, provided the correct grade of oil is used. Use heater plug position when starting from cold.

For example with a cold engine and an air temperature of 0°C (32°F) the key should be held in the heater plug position for 10 to 15 seconds. The time required for any set of circumstances will be found with experience. A red warning light will glow when the engine starter key is turned to the 'heater plug' position.

### Starting a cold engine
DO NOT use the accelerator pedal during the engine starting procedure; extra fuel for cold starting is automatically supplied by the injector pump.

**CAUTION:** The engine must not be run above fast idle until the oil pressure warning light goes off; this is to ensure that the engine bearings are receiving lubrication before being run at speed. This is **very important** on turbo-charged engines to ensure that the turbo-charger bearings are also receiving lubrication.

### Heater plug and starter switch – Fig. ST251
The heater plug and starter switch is combined with the steering column lock. The switch is key operated and has four positions.
'**0**': Key upright. All electrical circuits (except lights) switched off, steering lock engaged.
'**I**': With the key turned to the first position, the heater blower motor and accessories, such as radio, can be used, steering lock dis-engaged.
'**II**': Continue to turn the key against spring pressure. All electrical circuits switched on, heater plugs operating and amber warning light illuminated. Hold the key in this position for 10 to 15 seconds, then turn the key to the next position.
'**III**': Turn the key further against spring pressure to operate the starter motor. Release the key immediately the engine starts; the key will automatically return to the previous position. This is the 'run' mode with oil and charge lights and accessories.

*(continued)*

In cold weather, depress the clutch pedal while the starter motor is in operation to improve engine starting speed.

### Starter operation

 WARNING: Never, start or leave the engine running in an enclosed unventilated area. Exhaust fumes contain noxious substances which are harmful to health.

Do not operate the starter for longer than 10 seconds; switch off and wait 10 seconds before re-using the starter. If after a few attempts the engine fails to start, switch off and investigate the cause. Continued use of the starter will not only discharge the battery but may damage the starter.

### Stopping the engine – turbo-charged model

To avoid the possibility of inadequate lubrication of the turbo-charger, the following precaution must always be observed.

● Before stopping the engine, allow it to idle for 10 seconds to give time for the turbo-charger to slow down whilst oil pressure is available at the bearings.

● Switching the engine off too quickly could leave the turbine rotating at several thousand revolutions per minute without oil pressure.

### Starting a warm engine

DO NOT operate the accelerator pedal during the engine starting procedure. Turn the starter key to the engine start position. Release the key immediately the engine starts.

### Precautions for cold weather protection

The following recommendations should be considered to minimize difficulties associated with cold weather fuel problems.

● Ensure 'winter' grade fuel is used. Filling stations should automatically change to this fuel during winter.

● Renew the main fuel filter element at the recommended intervals.

● Maintain the state of charge of the battery in a satisfactory condition.

● Follow the starting procedures stated.

The use by operators, of paraffin (kerosene) as a diesel fuel additive, is illegal in the U.K. and the use of petrol as a fuel in a diesel engine is highly dangerous.

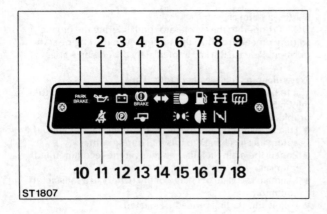

ST1807

**Key to warning light panel – Fig. ST1807**

| | | |
|---|---|---|
| 1. | Park brake (Australia only) | Red |
| 2. | Oil pressure | Red |
| 3. | Ignition | Red |
| 4. | Brake circuit | Red |
| 5. | Direction indicators | Green |
| 6. | Main beam | Blue |
| 7. | Low fuel | Amber |
| 8. | Differential lock | Amber |
| 9. | Heated rear window | Amber |
| 10. | Not used | |
| 11. | Seat belt warning (Saudi Arabia) | Red |
| 12. | Park brake — option | Red |
| 13. | Trailer — option | Green |
| 14. | Not used | |
| 15. | Side lights on | Green |
| 16. | Rear fog — option | Amber |
| 17. | Cold start | Amber |
| 18. | Not used | |

## Oil pressure warning light

The red warning light illustrated must glow when the ignition is switched on.

## Ignition warning light

The red ignition warning light illustrated should glow when the ignition is switched on.

NOTE: Ignition and oil warning lights should be checked when starting the vehicle from cold; they should light up immediately the ignition is switched on and extinguish when the engine is running. The warning lights may flicker when the engine is running at idling speed but provided they fade out as the engine speed increases, the charging rate and oil pressure are satisfactory. If the oil pressure warning light comes on during normal running, the vehicle should be stopped immediately and the cause investigated.
The ignition warning light is connected in series with the alternator field circuit. Bulb failure would prevent the alternator charging, therefore the bulb should be checked before suspecting an alternator fault. A failed bulb should be changed with the minimum of delay otherwise the battery will become discharged.

## Brake circuit check warning light

This red warning light is most important and is arranged to warn you if there is a fluid leakage from either the front or rear braking system when the engine is running. If leakage occurs the warning light will come on when brakes are applied. The brake circuit warning

light will operate momentarily when the starter is actuated. This confirms that the warning circuit is functioning correctly. If the light comes on during normal running or braking, the vehicle should be stopped immediately and the cause investigated.

## Direction indicator arrows

Both direction indicator arrows flash in conjunction with the direction indicator lamps, when operated by the switch on the steering column.
If the direction indicator arrows do not operate as described, there may be a bulb failure in the warning lamp panel or one of the direction indicator lamps.

## Main beam warning light

The blue light glows when the headlamp main beams are in use. Its purpose is to remind you to dip the headlamps when entering a brightly lit area, or when approaching other traffic.
The warning light will also glow when the headlamp flasher switch is used.

## Fuel level warning light

The amber warning light will be illuminated when there is approximately 9 litres (2 gallons) left in the fuel tank. The light will remain on until the fuel supply is replenished. Intermittent flashing may occur when cornering, etc. before the fuel level drops below two gallons. If a diesel model is allowed to run out of fuel, the fuel system must be primed when the tank is replenished.

**Differential lock warning light**

The amber warning light will be illuminated when the gearbox differential lock control is operated.

**Heated rear screen warning light**

The amber warning light will be illuminated when the heated rear screen switch is in the ON position, acting as a reminder to the driver that the switch and heated rear screen are switched ON.

**Trailer warning light**

The trailer warning light is operative when a trailer is connected to the vehicle via a seven-pin socket (optional equipment). It will flash in conjunction with the vehicle indicator warning lights, thus ensuring that the trailer indicator lamps are functioning correctly. In the event of an indicator bulb failure on the trailer, the warning light will flash once only and then remain extinguished. Where a trailer is not used or connected, the trailer warning light will only operate when the hazard warning system is in use.

**Side lights warning light**

The green warning light (with symbol) will be illuminated when the side lights are switched on.

**Rear fog guard lamps warning light**

The amber warning light will be illuminated when the rear fog guard lamps are operating.

**Cold start warning light**

**— Petrol models**
When the cold start control is pulled out, an amber warning light with this symbol is illuminated to remind the driver that the cold start control is still out and should be returned to the 'off' position as soon as possible, consistent with even running.

**— Diesel models**
On diesel models the amber warning light will glow when the engine starter key is turned to the heater plugs 'on' position. The light will go out as soon as the engine is started. If the light remains on with the engine running there is a fault that should be investigated. When operating in ambient temperatures of below −28°C, the use of a coolant heater is recommended.

**Hazard warning light**

When the hazard warning light switch is pressed at the lower end, all four flasher lights operate simultaneously. The red warning light (with triangular symbol) in the switch will flash in conjunction with the exterior flasher lights.
Use the hazard warning system to warn following or oncoming traffic of any hazard, that is, breakdown on fast road, or an accident to your own or other vehicles.

**Windscreen wiper switch and screen wash – Fig. ST017**
The windscreen wiper switch has five positions and is only operative when the ignition is switched on.
(A)    Switch in upper position: fast-speed wiper.
(B)    Switch in second position: slow-speed wiper.
(C)    Switch in third position: wipers off.
(D)    Switch in lowest position: 'flick-wipe' position: wipers will operate at slow speed until switch is released.
(E)    Switch pressed in: screen wash position. Hold the switch until sufficient water is ejected on to the screen, then release. This can be done with the wiper switch on or off.

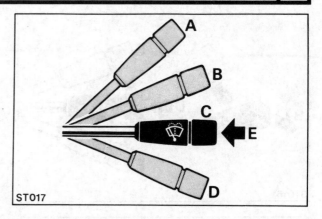

ST017

**Headlamp wash – Fig. ST266**
If the headlamp washer facility is fitted (optional), this will operate in conjunction with the windscreen washer when the headlamps are switched on in the dipped position.
The headlamp washer jet units (1) are fitted on the front bumper, one in front of each headlamp. The jet direction can be adjusted with the aid of a needle inserted into the orifice (2) which can also be cleared with a fine needle or wire when necessary.

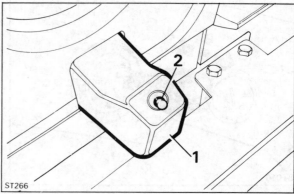

ST266

15

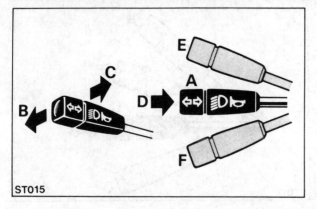

ST015

**Headlamp dipper switch, combining direction indicators, horn and headlamp flasher – Fig. ST015**

The switch has six positions:

(A)  Switch in central position: dipped headlamps.
(B)  Switch pushed away from driver: main beam.
(C)  Switch pulled towards driver: headlamp flash. The headlamps can be flashed at any time, irrespective of other switch positions.
(D)  Press dipper switch knob inwards to operate horn.
(E)  Move switch to upper position to indicate a right-hand turn.
(F)  Move switch to lower position to indicate a left-hand turn.

**Main light switch – Fig. ST016**

The main light switch has three positions:

(A)  Switch pulled towards driver: all lamps off.
(B)  Switch in centre position: side lamps on. (U.K. only: side lamps and dim dip headlamps on).
(C)  Switch pushed away from driver: side and headlamps on.

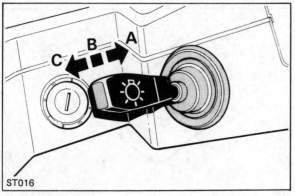

ST016

**Rear fog guard lamps switch (option) – Fig. ST018**
The switch has two positions and can be operated with or without the ignition on but is effective only with the headlamps on in the dipped position.
(A)   Switch pulled towards driver: fog lamp off.
(B)   Switch pushed away from driver: fog lamp on.

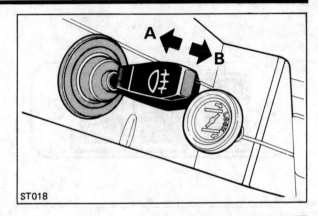

ST018

**Engine hand throttle (optional) – Fig. ST036**
This control will be found useful in conjunction with power take-off equipment and is used to over-ride the accelerator pedal linkage and set the throttle. This is suitable for all installations where precise speed control is not required, and where the engine load is light or relatively constant.
Pull the control out and twist it to lock it in the required position.
Operation of the accelerator will over-ride the hand throttle setting when increasing the engine speed. When the accelerator is released, the engine will return to the speed set by the hand throttle.
Before normal road driving is contemplated, check and ensure that the hand throttle is pushed fully down to the closed position. DO NOT use the hand throttle for motorway cruising.
Because the hand throttle is used to run the engine under load with the vehicle stationary, it may be necessary to fit an engine oil cooler system when used in hot climates.

**NOTE:** Always release the locking mechanism before returning the control to the 'OFF' position.

ST036

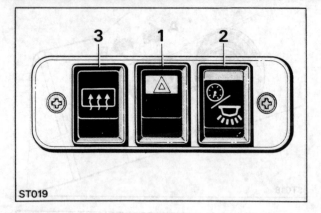

ST019

**Fig. ST019**
**Hazard warning switch**
1. The switch has a rocker action and the following positions:
   (a) Press the upper end of the switch: hazard warning system off.
   (b) Press the lower end of the switch: all flasher lights operate simultaneously.
   Use the hazard warning system to warn following or oncoming traffic of any hazard, that is, breakdown on fast road, or an accident to your own or other vehicles.

**Panel and interior light switch**
2. The switch has a rocker action and the following positions:
   (a) Press the upper end of the switch: panel lights on.
   (b) Return the switch to the centre position: lights off.
   (c) Press the lower end of the switch: interior lights on (if fitted).
   The panel lights are operative only with the main light switch at 'side' or 'head' position.

**Heated rear screen switch (when fitted)**
3. The switch has a rocker action and the following positions:
   (a) Press the upper end of the switch: heated rear screen switched off.
   (b) Press the lower end of the switch: to operate the rear screen demisting heater. This position will only be operative whilst the ignition is switched on. The integral warning lamp is lit when the switch is in the ON position, acting as a reminder to the driver that the switch and screen are on.

**Fig. ST020**

**Speedometer**

1. The speedometer incorporates a total mileage indicator. Speedometers with trip mileage indicators are available as optional equipment and have a trip reset button fitted.

**Speedometer trip setting**

2. Reset trip back to zero by pushing the small black knob on the front of the speedometer.

**Fuel level indicator**

3. The fuel level indicator shows the approximate contents of the tank.

**Coolant temperature indicator**

4. Under normal running conditions the temperature indicator needle should register in the black band. If the needle moves to the red band during normal running, the vehicle should be stopped and the cause investigated.

The design of the fuel level and water temperature indicators ensures that the needle does not fluctuate, but there is a time lag of a few seconds before they register after the ignition, or electrical services, are switched on.

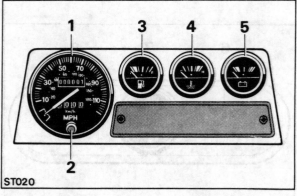

ST020

**Voltmeter (option)**

5. The voltmeter measures the vehicle system voltage. With the engine running above idling speed the indicator should register within the black central band. A reading above this in the high red band which continues after 10 minutes running is too high and should be investigated. A reading in the low red band with the engine running at high idle speed, with no electrical loads switched on, after 10 minutes is too low and should be investigated.

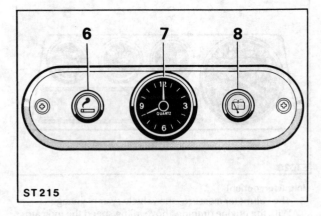

ST 215

**Fig. ST215**

**Cigar lighter (option)**

6. The cigar lighter is operated by pushing the extended knob inwards to heat the element. When a predetermined temperature is reached, the knob will eject from the heat position, permitting the lighter to be withdrawn for use.
   A small pilot lamp is incorporated within the socket surround to facilitate replacement of the element during darkness. The pilot lamp bulb is automatically lit when the vehicle sidelights are on.

**Clock (option)**

7. The hands of the electrically operated clock may be set by pushing in and turning the black knob in the centre of the face.

**Rear screen wiper and washer switch (option)**

8. The rear screen washer/wiper switch is only operative in the ignition or accessory position on the ignition switch.
   (a) Rotate the switch to the right to activate the rear screen wiper.
   (b) To wash the rear screen, press the spring loaded switch knob until sufficient water is on the rear screen. Releasing the knob will shut off the rear screen wash water. This operation may be carried out with the screen wiper switch ON or OFF.

**Fig. ST021**

**Oil pressure gauge (option)**

9.  Under normal running the oil pressure indicator should show the following pressure:
    **4-cylinder petrol and diesel models —**
       2,5 to 4,5 kgf/cm² (35 to 65 lbf/in²) 240 to 440 kPa
    **V8-cylinder petrol models —**
       2,1 to 2,8 kgf/cm² (30 to 40 lbf/in²) 200 to 275 kPa
    The needle may drop below these figures when the engine is idling but providing the oil pressure rises to within the specified figures immediately the engine speed is increased, the oil pressure can be considered to be satisfactory.
    If the needle moves to the zero position during normal running the vehicle should be stopped immediately and the cause investigated.

**Oil temperature gauge (option)**

10. The oil temperature gauge provides a continuous indication of the oil temperature. When the engine oil reaches its normal operating temperature, the gauge indicator needle should register in the mid-way area. Should the needle travel to the 'H' (hot) red block during normal running, the vehicle must be stopped and the cause investigated.

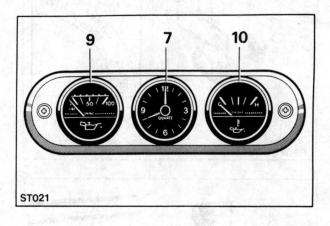

ST021

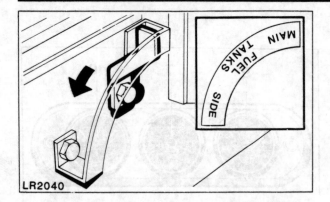

LR2040

**Fig. LR2040**

**Fuel tank changeover switch — One Ten models**
If the vehicle is fitted with an extra fuel tank (option) a combined changeover tap and switch is located on the heelboard. Movement of the tap lever brings into use either the rear or the side tank, and switches the fuel level indicator to show the approximate contents of the tank in use. When the lever is in the horizontal position the side tank is in use, in the vertical position the main tank is in use.

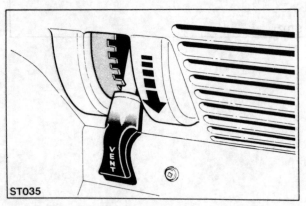

ST035

**Windscreen ventilators – Fig. ST035**
The two ventilators in the windscreen frame may be opened independently by pushing the lever downwards until each ventilator is open to the desired position. Use of the ventilators will be found advantageous when traversing dusty roads, as they greatly reduce the amount of dust sucked into the vehicle from the rear.

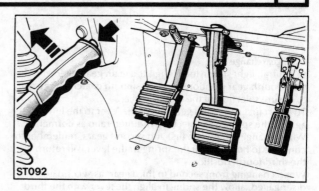

ST092

## Transmission handbrake – Fig. ST092

1. A drum-type handbrake, well protected from dirt and water, operates directly on the transfer box rear output shaft and is designed for parking use only.
   The brake is applied by pulling back the lever. To release, pull the lever slightly back, depress and hold the release button while pushing the lever down to the limit of its travel.

 **WARNING: DO NOT apply the handbrake while the vehicle is in motion as this could result in loss of vehicle control and damage to the transmission.**

## Pedals – Fig. ST092

2. Brake, clutch and accelerator pedals are the pendant type and function in the normal way. The brake and clutch operate hydraulically, with servo assistance for the brakes. The accelerator pedal has a mechanical linkage. To avoid needless wear of the clutch withdrawal mechanism do not rest the foot on the clutch pedal while driving.

## Gearbox controls and ranges – Fig. ST025

The main gearbox of the Land Rover is augmented by a two-speed transfer box giving high and low ranges. Therefore the five-speed manual gearbox used in conjunction with the transfer gearing produces ten forward and two reverse ratios.

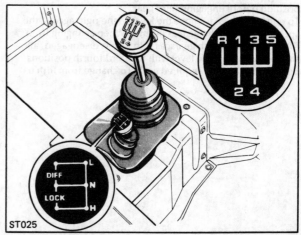

ST025

*(continued)*

**Main gearchange lever**

In neutral, light spring loads align the main gear lever with the third/fourth gear positions to assist smooth gearchanging and to ensure selection of the required gear.

To select first or second gear, move the lever to the left against the spring and select the required ratio as normal. When changing between first and second gears, remember to continue to hold against the spring or the lever will return to the third/fourth position.

When changing from second to third gear, as second gear is disengaged, allow the spring to align the lever with the third position before engaging third gear.

To engage fifth gear, move the lever to the right against the spring and select the gear as normal. When changing from fifth to third or fourth gears, as fifth gear is disengaged, allow the spring to align the lever with the third/fourth positions before engaging the required gear. To change from fifth to second or first gear, allow the lever to return to the third/fourth position and move the lever towards the left against the spring as already described. Note that fifth gear is designed to reduce engine speed and thus improve fuel economy when cruising. Ensure that while it is in use the engine runs easily without labouring, otherwise use a lower gear.

Reverse is protected against inadvertent selection by an additional 'knock-over' spring load. To engage reverse, strike the lever as far as possible towards the left using the palm of the hand and move it forward to engage the gear. To disengage, pull the lever rearwards and allow the spring load to return it to its normal position in neutral.

It is recommended that, before driving away for the first time, the driver becomes familiar with the operation of the gear change by changing up and down through all ratios several times.

**Combined transfer gear and centre differential lock lever – Fig. ST026**

The transfer gear lever controls the selection of the high or low gear ranges and the engagement of the centre differential lock. The lever, which is located immediately behind the main gear lever, has the following positions:

**Central right.** Transfer box in neutral, Position **N**, centre differential unlocked. In this position drive cannot be transmitted to the road wheels regardless of the position of the main gear selector. Use this position for winching or power take-off (pto) and when being towed.

**Fully forward and right, Position L.** Transfer gearbox low range engaged.

**Fully forward and left, Position L.** Transfer gearbox low range engaged AND centre differential locked (warning light illuminated).

**Fully rearwards and right, Position H.** Transfer gearbox high range engaged. This position is used for normal driving.

**Fully rearwards and left, Position H.** Transfer gearbox high range engaged AND centre differential locked (warning light illuminated).

**Centre left.** Transfer box in neutral, Position **N**, centre differential locked. (This position should not be used).

**NOTE:** On a new vehicle, transfer gear lever operation may be a little stiff until the gearbox is 'run-in'.

**Use of the transfer gear lever**

**CAUTION:** Changing from high (**H**) to low (**L**), **should only be attempted when the vehicle is stationary.** Depress the clutch pedal and push the lever fully forward, release the clutch. Should there be any hesitation in the gear engaging, do not

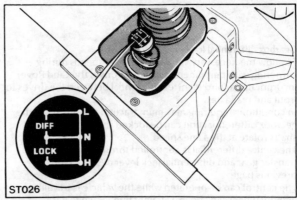

ST026

force the lever. With the engine running, engage a gear with the main gear lever and release the clutch momentarily, then return the main gear lever to neutral and try the transfer control again.

Changes from low (**L**) to high (**H**) can easily be made as follows without stopping the vehicle.

Depress the clutch pedal and release the accelerator pedal as for a normal gearchange. Move the transfer lever into neutral. Release the clutch pedal for 3 seconds. Depress the clutch pedal and move the transfer lever firmly to the 'high' (**H**) position. Then move the main gear lever to second gear and release the clutch pedal while depressing the accelerator to take up the drive smoothly. As the vehicle accelerates, change gear in the main gearbox in the normal way.

This operation can be carried out smoothly and quickly after a little practice. Proper use of the gearbox range will ensure optimum efficiency and transmission component life.

**Gearbox differential lock**

To allow the necessary variation of wheel speeds during cornering with permanent four-wheel drive, the Land Rover incorporates a third (centre) differential between the drives to front and rear axles.

In conditions requiring maximum traction to both axles, the gearbox differential unit can be locked so that both output shafts rotate at the same speed.

The centre differential is controlled through the combined transfer gear and differential lock lever described on the previous page.

The control can be operated while the vehicle is travelling without wheel slip and in a straight line, or while it is stationary. The differential should be locked before slippery or doubtful surface conditions are encountered.

**CAUTION:** Engagement of the lock with one or more wheels slipping will cause damage to the transmission.

Under certain conditions a slight delay may be experienced before the differential becomes locked, with subsequent warning light illumination. This delay is a built-in safety precaution and ensures that gears are correctly aligned before differential locking occurs.

On disengagement of the lock there may be a short delay before the warning light goes out indicating differential unlocked. If the warning light remains on, this indicates that the transmission is 'wound-up'. The vehicle must be stopped and reversed for a few metres to 'unwind' the transmission; the warning light will then be extinguished and the vehicle can proceed.

**Front seats – Fig. ST008**

**Fore and aft adjustment**

1. Lift the bar at the front of the seat and slide the seat to the required position. Release the bar and ensure the seat guide catches have located the seat.

**Back rest angle adjustment (option)**

2. Ease the body from the back rest and lift the locking handle. Apply body pressure to move the back rest to the required rake, then press the handle down to lock. The back rest return is spring assisted.

**Height adjustment**

3. Remove the seat cushion by lifting it from its retaining clip. Adjust to required height by sliding each end of the sprung rod (situated underneath the cushion) into the desired height adjustment hole. Replace the cushion by inserting the rear location pegs at the base of the cushion. Firmly press the cushion into the guide retaining clip.

**Head restraints**

4. Head restraints can be fitted to seats with adjustable back rests on all models except the truck cab.

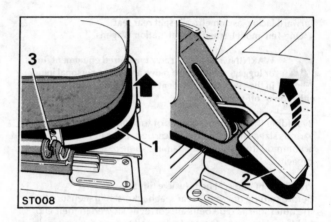

ST008

**Forward facing, asymmetrically split rear seat
– 'One Ten' model, 9 and 10 Seat Station Wagons**

 **WARNING: DO NOT carry unsecured equipment, tools or luggage which could move and cause personal injury in the event of an accident or emergency manoeuvre either on or off-road.**

By folding the separate sections of the split rear seat, loads of various sizes and shape can be carried. Long items can be accommodated while still retaining some rear seating capacity.

**Protection of rear seat belts (where fitted)**
Before folding down the rear seat backrests, first ensure that the outer inertia type belts are correctly stowed in their clip holders. Also, keep the centre lap belt fastened when not in use. To avoid damage to the inner sections of the inertia type belts and the centre lap belt mounted on the floor behind the rear seat, pass the four belts between the bottom of the seat backs and the seat to the rear floor.

Before erecting the rear seat, ensure that all inner seat belts are extended rearwards to prevent them from being trapped beneath the seatbase.

If the vehicle payload is likely to damage or chafe the belts in the rear floor area, they should be removed temporarily. In this event, unhook the belts from their respective floor mounted brackets by holding open the spring loaded safety catch. After reconnection, ensure that the safety catch returns to the closed position.

**NOTE:** Australia only — The belts are bolted to the rear floor.

**Carrying loads in rear passenger area**
To convert the rear passenger area for load carrying, fold up the base of each inward facing seat and hook the retaining strap over the backrest frame.

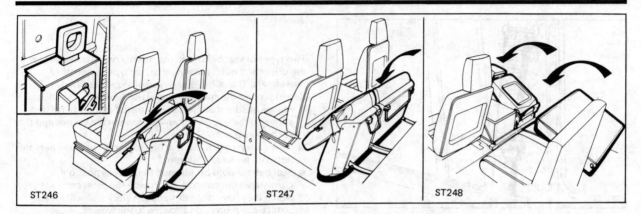

ST246  ST247  ST248

**Carrying bulky loads**

Slide the front seats forward sufficiently to allow the rear seat backrests to be folded.

● Pull up the ring-type release handle (inset Fig. ST246) located on the window ledge behind each backrest. Fold the appropriate section of the backrest and tip the folded seat forward.

● When returning the seats to the normal position, check that the rear support legs are pulled fully back, and that both backrest locks are correctly engaged.

● To accommodate extra long loads, fold the appropriate section of the backrest forward and incline the front passenger seat fully forward — Fig. ST248.

**Maximum floor space – Fig. ST247**

Slide the front seats forward sufficiently to allow the rear seat backrests to be folded.

● Pull the release handles upward, fold the rear seat backrests and tip both sections of the seat assembly forward.

**Roof racks**

Land Rover vehicles incorporating roofs with aluminium cantrails (rain water gutter) require the use of an approved roof rack. Information concerning suitable roof racks is available through the Land Rover parts service. These should be fitted very carefully following the manufacturers' instructions.

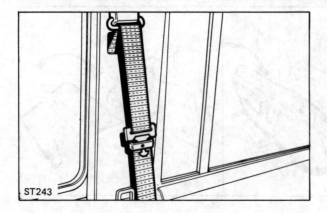

ST243

Two types of seat belt are in use, inertia reel (automatic) for the driver and outer passenger(s), lap type for all other passengers. The number and type of seat belts fitted is dependent on the specification of the vehicle.

- Always ensure that the belt is lying flat and is not twisted either on the wearer's body or between the wearer and the anchorage point.
- Never attempt to use a seat belt for more than one person, not even for small children.
- Seat belts should be adjusted as firmly as possible, consistent with comfort, to provide the protection for which they have been designed. A slack belt will greatly reduce the protection afforded to the wearer.

**General**

 **WARNING: Seat belts are designed to bear upon the bony structure of the body, and should be worn low across the front of the pelvis, or the pelvis, chest and shoulders, as applicable; wearing the lap section of the belt across the abdominal area must be avoided.**

All seat belts must be fitted to the anchorage points provided at both the drivers and passenger's position to comply with the United Kingdom or other territorial legal requirements.

- In your interests, always use the seat belt provided, even for the shortest journeys. Alterations and additions must NOT be made to any type of seat belt fitted to this vehicle.

**Inertia reel seat belt – Fig. ST088**
To fasten, draw the tongue of the belt over the shoulder and
across the chest, then push it into the engagement/release
slot. A positive click indicates that the belt is safely locked.
● To release, press the release button which will
automatically dis-engage the buckle; this allows the belt to
retract. Position the moveable clip as high as possible so
that the tongue is accessible when the belt is next required.

**Lap seat belt – Fig. ST242**
The lap belt is fastened and released in the same way as the
inertia reel type.
● To adjust, slide the adjuster along the belt and feed the
webbing through the buckle until the belt is comfortably
tight.
● When not in use, lap belts should be fastened.

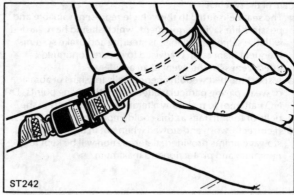

**Testing inertia reel type seat belt**

 **WARNING: This test must be carried out under safe road conditions, that is, level dry road with no following or oncoming traffic.**

With the seat belt is use, drive the vehicle at 8 kph (5 mph) and brake sharply. The automatic locking device should operate and lock the belt. It is essential that the driver and passenger are sitting in a normal relaxed position when making the test. The retarding effect of the braking must not be anticipated.

**Care of the seat belt**

● The seat belts fitted to this vehicle represent valuable and possible life saving equipment, which should be regarded with the same importance as steering and brake systems. Frequent inspection is advised to ensure continued effectiveness in the event of an accident.

● Inspect the belt webbing periodically for signs of abrasion or wear, paying particular attention to the fixing points. DO NOT attempt to make any alterations or additions to the belts or their fixings as this could impair their efficiency.

● If correctly worn and stowed when not in use, on the stowage points provided, deterioration will be kept to a minimum and protection to a maximum.

● Seat belt assemblies must be replaced if the vehicle has been involved in an accident or if upon inspection, there is evidence of cutting or fraying of the webbing, incorrect buckle or tongue locking function; and/or any damage to the buckle stalk cabling.

**Seat belt cleaning**

DO NOT attempt to bleach the belt webbing or re-dye it. If the webbing becomes soiled: sponge with warm water using a non-detergent soap and allow to dry naturally. DO NOT use caustic soap, chemical cleaners or detergents for cleaning; do not dry with artificial heat or by direct exposure to the sun.

**Child restraint upper anchorages –**
forward facing rear seats. Australian design rule number 34A.

 **WARNING: Child restraint anchorages are designed to withstand only those loads imposed by correctly fitted child restraints.**
**Under no circumstances are they to be used for adult seat belts or harnesses.**
**Child restraints are designed to bear upon the bony structure of the body as are the seat belts for adults.**

The method of fixing the upper anchorage fittings, dimensions of spacers required and length of bolts are shown opposite. The child restraint must be fitted in accordance with the seat belt manufacturer's instructions.

1. Spacer.
2. Upper anchorage fitting.
3. Plain washer.
4. Securing bolt, minimum length.
5. Spacer dimensions.
6. $\frac{7}{16}$-18 UNC-2B threaded tube welded into mounting bar.
7. Seat belt mounting bar.
8. Mounting bar fixing to vehicle.

**NOTE:** Items 1 to 4 supplied by seat belt manufacturers.

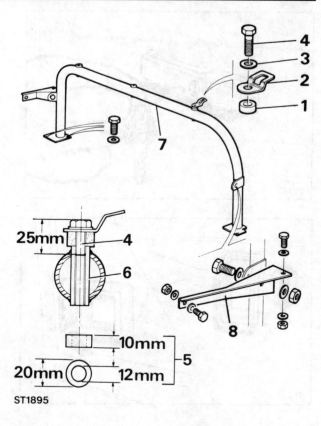

ST1895

33

ST233

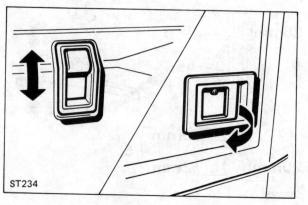

ST234

### Door lock operation

#### From outside – Fig. ST233

To lock a front door, turn the key rearward a quarter of a turn, return the key to the vertical position and remove it.
- To lock a front door without using a key (Take care not to leave the keys inside the vehicle).
- Hold the external release button in and depress the interior locking button, release the external button and close the door.
- To unlock a front door, insert the key and turn it forward a quarter of a turn, return the key to the vertical position and remove it.
- To lock a rear side door — One Ten Station Wagons — push down the interior locking button. This can be done with door open or closed.

#### From inside – Fig. ST234

To lock any door, push down the interior locking button.
To unlock any door, push up the interior locking button.

#### Windows (side doors)

To raise or lower the door windows turn the handle either to the right or left in a circular motion, as required.

**Child proof locking – 'One Ten' Station Wagon rear side doors –
Fig. ST235**
Each rear side door is fitted with a child proof lock. Move the
setting lever down to prevent the door being opened from
inside the vehicle.

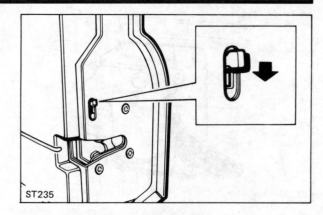

ST235

**Rear door – Hard Top models and Station Wagons**

**From outside**
To unlock the door, insert the key and turn it anti-clockwise a
quarter of a turn, return the key to the vertical position and
remove it.

- To open the door, simply lift the outside handle. When the
  door is fully opened, a catch automatically locks the check
  strap and holds the door in the open position.
- To close the door, push the release lever for the check
  strap towards the door hinge and close the door —
  Fig. ST236.
- To lock the door, insert the key and turn it clockwise a
  quarter of a turn, return the key to the vertical position and
  remove it.

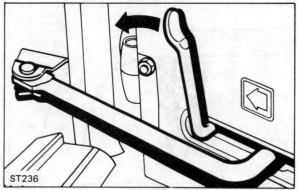

ST236

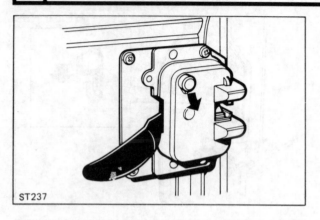
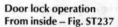

ST237

### Door lock operation
### From inside – Fig. ST237
To unlock the door move the knob on the lock case upward.
Open the door using the inside handle.
● To close the door, push the release lever for the check
   strap towards the door hinge and close the door.
● To lock the door, move the knob on the lock case
   downward before, or after, closing the door.

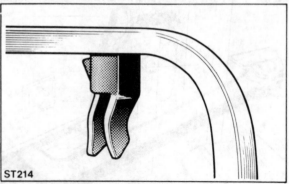

ST214

### Rear side windows – sliding type (option) – Fig. ST214
The forward section of the sliding type rear side windows can
be opened as required for rear passenger ventilation. Each
window is controlled by a single catch. To open, press the
catch tongues together, slide the window to the desired
aperture position and release the catch which will
automatically lock the windows in position.

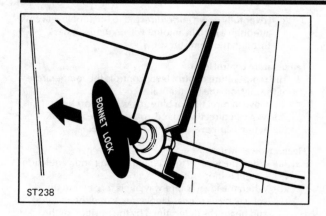

ST238

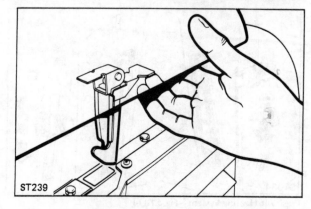

ST239

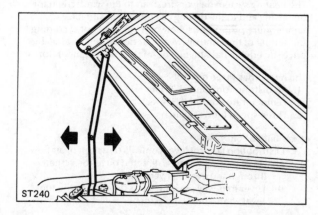

ST240

**Bonnet**

 **WARNING: If the spare wheel is fitted on the bonnet, it will be heavy to lift; DO NOT allow it to drop.**

The bonnet release is located under the dash, to the right of the gearbox tunnel — Fig. ST238
- To release the bonnet catch, pull the bonnet release handle.
- From outside the front of the vehicle, lift the safety catch lever and raise the bonnet — Fig. ST239. Pull the support stay forward to secure the bonnet in the open position — Fig. ST240.
- To close, raise the bonnet slightly, support it while pushing back the support stay, and lower the bonnet. Press down on the forward edge of the bonnet with the hands to engage the lock.

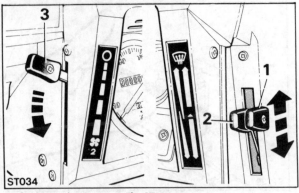

ST034

## Fresh Air/Heating System – Fig. ST034

The heating system delivers fresh air to the windscreen for demisting and to the driving cab interior in variable temperature proportions, between cold and hot according to the setting of the controls. Warm or hot air will be available once the engine has attained normal working temperatures.

The heater has three controls:
1. Air distribution control (RH lever).
2. Temperature control (LH lever).
3. Three speed blower switch.

### Distribution control lever

1. Distribution control lever controls direction of air flow.
   (a) Lever fully up, all air is directed on to the screen through the demister vents.
   (b) Lever mid-way position, air is directed to the foot level vents and to the screen.

(c) Lever fully down, air is directed to the foot level vents although a certain amount will continue to pass through the demister vents.

### Temperature control lever

2. The temperature control lever controls the temperature of the air from the heater unit.
   (a) Move in direction of blue arrow to cut off heat.
   (b) Move in direction of red arrow to increase heat.
   (c) Action is progressive between the two.

### Heater blower switch

3. (a) When the lever is fully up the heating and ventilation system is inoperative.
   (b) When the lever is mid-way, air is forced into the vehicle by its forward movement, and then routed and heated as determined by the position of the distribution control and the heat control. When the vehicle is stationary, the system is inoperative.
   (c) The blower motor will only operate with the engine running or the starter key turned to the first position. Move the lever down to the first or second stop, this will give slow or fast blower motor speed to boost the air flow into the vehicle. Air is routed and heated as determined by the position of the distribution control and the heat control.

### Demisting

Mist often forms on windows when the humidity is very high.

**NOTE:** For maximum demist effectiveness, the fresh air supply should be used. Place the right-hand lever (1) in the top position. Place the left-hand lever (2) in the bottom position and the blower switch (3) in the mid-way position.

## Using the air conditioning – Fig. ST227
Set the controls as follows:
1. Temperature control to blue zone — Fully down (cold).
2. Recirculation control — Fully up (recirculation).
3. Distribution control to mid position (fascia).
4. Fan control set between positions 1 to 4 as desired, to regulate the volume of airflow desired.
5. Air conditioning control pushed in to illuminate the pushbutton legend.

When the temperature inside the vehicle becomes comfortable, move the temperature control up slightly. This will prevent the evaporator cooling coils from becoming too cold and freezing up.

### Rapid cooling
Open a window.
Move the fan control to position 4.
Move the temperature control down to the coldest position.
Set the distribution control to mid position (fascia).
Recirculation control set to recirculation (fully up).
After driving for several minutes, the hot air inside the vehicle will be expelled. Close the window, move the temperature control down slightly and adjust the fan speed as desired.

### Demisting
Mist often forms on windows when the humidity is very high. To remove the mist, move the temperature and fan controls to their low positions. If the interior temperature is too low, use the heater in conjunction with the air conditioning. It is not necessary to use the system continuously, only when misting persists.

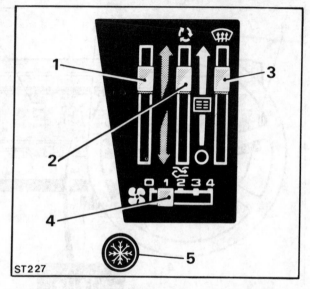

ST227

**NOTE:** For maximum demist effictiveness, use the fresh air supply (described earlier). Used in conjunction with the maximum heater setting, the air conditioning system will produce an air drying effect which will assist demisting.

### Heating
During cold weather the fan can be used to circulate warm air from the heater.
Move both the fan and temperature controls to the desired setting.

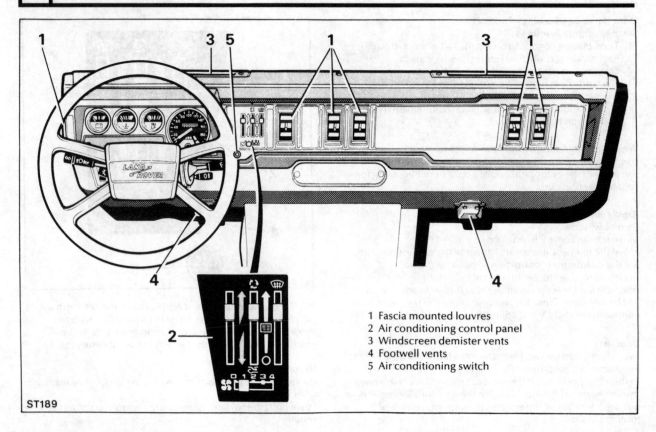

1 Fascia mounted louvres
2 Air conditioning control panel
3 Windscreen demister vents
4 Footwell vents
5 Air conditioning switch

ST189

### Fascia-mounted louvres

The six fascia-mounted louvres can be set to blow cooled, fresh or recirculated air, the vanes may be opened and adjusted to control the direction of airflow.

### Fan control

The fan control should be adjusted to regulate the volume of air required.

### Air conditioning control

The air conditioning pushbutton control is pressed to switch on the air conditioning and is illuminated when operative.

### Temperature control

The temperature of air flowing from the footwell and windscreen may be regulated between cold (blue) and hot (red) by moving the control as required. For effective air conditioning, this control should be maintained in the cold (blue) position.

### Air distribution control

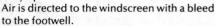

The distribution control has three positions.

(a) Fully up.
   Air is directed to the windscreen with a bleed to the footwell.
(b) Central position.
   This position is used to direct air from the fascia-mounted louvres, with a bleed to the footwell.
(c) Lower position.
   Air is directed to the footwells, although a certain amount will continue to flow through the demister vents to the windscreen.
   Any of the air distribution positions may be used in conjunction with the temperature fan and air conditioning controls.

### Recirculation/Fresh air control

The vehicle has a combined fresh air, or recirculating air system, designed to enable either system to be used separately. The air fed through the air distribution control can be either fresh air drawn from outside the vehicle or internally recirculated air. The recirculating heater is normally used in heavy traffic conditions to avoid obnoxious fumes entering the vehicle, also for rapid heat build up inside the vehicle during cold conditions. The recirculating control is used with air conditioning to achieve maximum cooling. It is also recommended that the recirculating control is used in dusty conditions to prevent dust entering the vehicle.

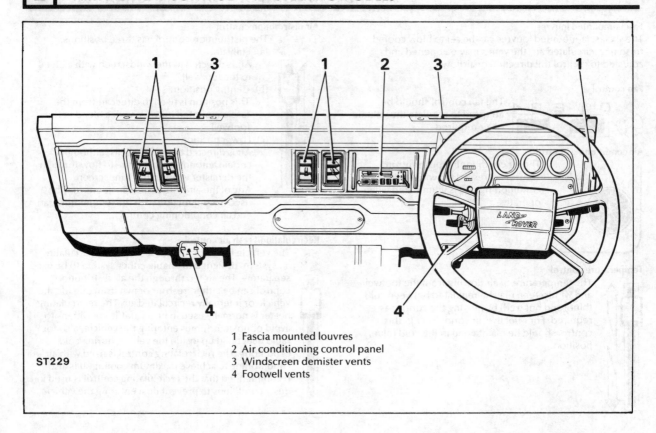

ST229

1 Fascia mounted louvres
2 Air conditioning control panel
3 Windscreen demister vents
4 Footwell vents

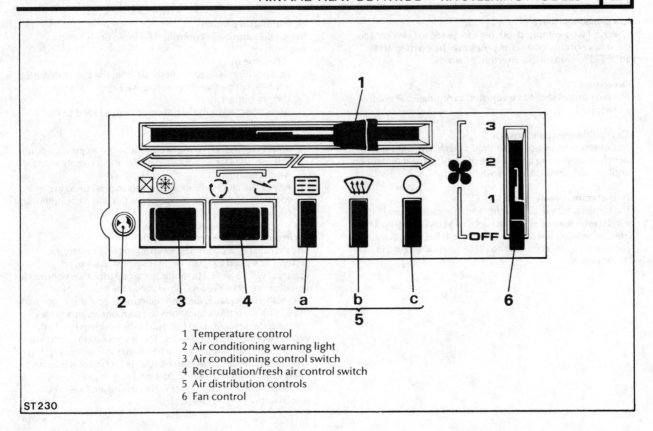

1 Temperature control
2 Air conditioning warning light
3 Air conditioning control switch
4 Recirculation/fresh air control switch
5 Air distribution controls
6 Fan control

ST 230

**Fascia-mounted louvres**

The five fascia-mounted louvres can be set to blow cooled, fresh or recirculated air, the vanes may be opened and adjusted to control the direction of airflow.

**Fan control**

The fan control should be adjusted to regulate the volume of air required.

**Air conditioning control switch**

To switch on the air conditioning, push in the right side of the switch. The warning light will be illuminated and remain on until the air conditioning is switched off.

**Temperature control**

The temperature of air flowing from the footwell and windscreen may be regulated between cold (blue) and hot (red) by moving the control as required. For effective air conditioning, this control should be maintained in the cold (blue) position.

**Air distribution controls**

The air distribution is controlled by three push button switches.

(a) LH button in

   This position is used to direct air from the fascia-mounted louvres, with a bleed to the footwell.

(b) Centre button in

   Air is directed to the windscreen with a bleed to the footwell.

(c) RH button in

   Air is directed to the footwells, although a certain amount will continue to flow through the demister vents to the windscreen.

   Any of the air distribution controls may be used in conjunction with the temperature fan and air conditioning controls.

**Recirculation/Fresh air control switch**

The vehicle has a combined fresh air, or recirculating air system, designed to enable either system to be used separately. The air fed through the air distribution control can be either fresh air drawn from outside the vehicle or internally recirculated air. The recirculating heater is normally used in heavy traffic conditions to avoid obnoxious fumes entering the vehicle, also for rapid heat build up inside the vehicle during cold conditions. The recirculation control is also with air conditioning to achieve maximum cooling. It is also recommended that the recirculating control is used in dusty conditions to prevent dust entering the vehicle.

Push the switch to the left for recirculating air. Press the switch to the right for fresh air.

## Using the air conditioning – Fig. ST232
Set the heater controls as follows:
1 Temperature control to blue zone — Fully left (cold).
2 Distribution control — Push in a to c as required.
3 Recirculation control — Push in left side of switch.
4 Fan control set between positions 1 to 3 as desired, to regulate the volume of airflow desired.
5 Air conditioning control — Push in right side of switch to switch air conditioning on and illuminate warning light.

When the temperature inside the vehicle becomes comfortable, slide the temperature control to the right slightly. This will prevent the evaporator cooling coils from becoming too cold and freezing up.

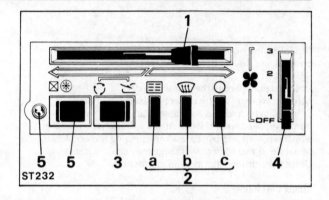

### Rapid cooling
Open a window.
Move the fan control to position 3.
Move the temperature control to the left to the coldest position.
Push in the distribution control, as required.
Recirculation control set to recirculation — Push in left side of switch.
After driving for several minutes, the hot air inside the vehicle will be expelled. Close the window, move the temperature control to the right slightly and adjust the fan speed as desired.

### Heating
During cold weather the fan can be used to circulate warm air from the heater. Move both the fan and temperature controls to the desired setting.

### Demisting
Mist often forms on windows when the humidity is very high. To remove the mist, move the temperature and fan controls to their low positions. If the interior temperature is too low, use the heater in conjunction with the air conditioning. It is not necessary to use the system continuously, only when misting persists.

**NOTE:** For maximum demist effectiveness, use the fresh air supply (described earlier). Used in conjunction with the maximum heater setting, the air conditioning system will produce an air drying effect which will assist demisting.

# 3 | DRIVING TECHNIQUES

The following notes are a general guide to the technique of driving Land Rovers over rough terrain.

### Match engine speed to the gear selected
Before traversing a difficult section, select low range differential locked and a suitable gear, which, for most purposes, second or third is satisfactory. Remain in this gear whilst crossing and use care when applying the accelerator pedal since a sudden power surge may cause wheel spin. Unlock the differential as soon as practical.

### Riding the clutch
Keep the foot away from the clutch pedal. The practice of resting the foot on the clutch pedal should be avoided. Apart from premature clutch wear a sudden bump could cause the pedal to be depressed too far, disengaging the drive, and causing the vehicle to go out of control.

### Braking
Keep the application of the brake pedal to a minimum. Braking on wet or muddy slopes can induce sliding and loss of control.

### Use of engine for braking
Before descending steep slopes, first gear low range with differential locked should be selected and the engine should be allowed to provide the braking. This it will do without assistance from the wheel brakes. Failure to adopt this procedure may result in loss of control.

### Driving on soft ground
Where conditions are soft, such as marsh ground or sand, reduced tyre pressures will increase the contact area of the tyres with the ground. This will help to improve traction and reduce the tendency to sink. Tyre pressures should be reinflated to the standard pressures when firm ground is reached.

### Rough rocky tracks
Although beaten rough tracks can be negotiated in normal drive, it is advisable to lock the differential if there is excessive suspension movement likely to induce wheel spin. As the track becomes rougher and more rocky, low range may be necessary to avoid slipping the clutch and to make the Land Rover easier to control. Do not hold the steering wheel with the fingers and thumbs inside the wheel. A sudden violent kick of the wheel could damage or even break the fingers. Grip the wheel on the outside of the rim when travelling across country.

## Climbing steep slopes

This will usually require the use of low range second or third gear with differential locked. Should the slope be slippery use the highest gear that the engine can manage without labouring and stalling.

If the vehicle fails to climb a slope but does not stall, the following procedure should be carried out. Hold the vehicle on the footbrake and engage reverse gear as quickly as possible. Release the brakes and allow vehicle to reverse down the slope whilst ensuring that both feet are clear of the brake and clutch pedals.

If the vehicle stalls on a slope, hold the Land Rover on the footbrake, engage reverse gear and remove the feet from both clutch and brake pedals. Start the engine whilst in gear and allow the Land Rover to reverse down the slope, using only the retardation effect of the engine for braking.

When back on level ground, or where forward traction can be regained, then a possible faster approach will overcome the inertia and the extra momentum will often enable the slope to be climbed.

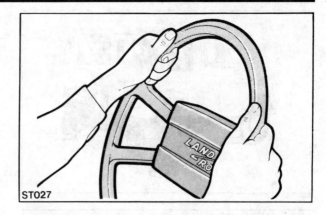

ST027

## Ground clearance

Be aware of the need to maintain ground clearance under the chassis and a clear approach and departure angle. Avoid existing deep wheel ruts, sudden changes in slope and obstacles which could interfere with the chassis.

## Rutted and existing wheel tracks

Generally the tendency is to over-steer the vehicle under these circumstances, resulting in the vehicle being driven on left- or right-hand lock in ruts. This should be avoided as it produces drag at the road wheels and can be dangerous, causing the vehicle to veer off the track the moment the front wheels reach level ground or find traction.

*(continued)*

ST 216

**Crossing a ditch – Fig. ST216**
Ditches should be crossed at an angle so that three wheels are kept in contact with the ground. If approached at right angles the two front wheels will drop into the ditch, effectively preventing forward or rearward movement.

ST096

**Traversing slopes – Fig. ST096**
Traversing a slope should be undertaken in the following way. Check that the ground is firm under all wheels and that it is not soft under the downhill side wheels. Also avoid the uppermost wheels climbing up over a rock or tree root, both of these situations could result in the vehicle rolling on to its side.

*(continued)*

**Negotiating a 'V' shaped gully – Fig. ST028**
This should be tackled with caution since steering up or down the gully walls could lead to the vehicle becoming trapped on the bank or on an obstacle such as a tree or rock.

**Crossing ridges and ditches – Fig. ST095**
Select a path so that the condition under each wheel is similar to that under the opposite wheel of the same axle. This principle should be applied both in avoiding dissimilar ground surfaces under opposite wheels and in assessing the correct angle of approach to an obstacle so as to avoid the wheels being lifted off the ground.

**Crossing over a ridge**
Approach a ridge at right angles so that both front wheels go over together. If approached at an angle, traction can be completely lost through diagonally opposite wheels leaving the ground.

49

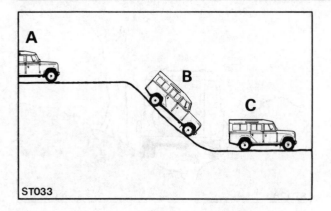

ST033

**NOTE:** The differential lock can be engaged or disengaged at any speed providing the road wheels are not spinning at different revolutions. For example, in slippery conditions if one wheel is spinning, ease off the accelerator before engagement.

### After cross country driving
If the tyre pressures have been reduced, they MUST be restored to the normal recommended pressures as soon as reasonable road conditions or hard ground is reached.

⚠️ **WARNING: Before rejoining the highway (public metalled roads) or driving the vehicle at speeds above 40 km/h (25 mph), wheels and tyres must be inspected for damage, after cleaning off any mud. Do not forget the inside faces.**

### Vehicle recovery
Should the vehicle become immobile due to loss of wheel grip, the following hints could be of value:
(a)  Avoid prolonged wheel spin; this will only make matters worse.
(b)  Try to remove obstacles rather than force the vehicle to cross them.
(c)  If the ground is very soft, reduce tyre pressures if this has not previously been done.
(d)  Clear clogged tyre treads.
(e)  Reverse as far as possible, then the momentum reached in going forward again may get the vehicle over the obstacle.
(f)  Brushwood, sacking, or any similar 'mat' material placed in front of the tyres will help in producing tyre grip.
(g)  If possible, jack up the vehicle and place material under the wheels. Great care must be taken when doing this to avoid personal injury.

### Descending steep slopes – Fig. ST033
Stop the vehicle at least a vehicles length before the slope and engage first gear, low range with the differential locked. Check gear engagement before moving off. Do not touch the brake or clutch during the descent — the engine will limit the speed, and the vehicle will remain perfectly under control while the front wheels are turning. If the vehicle begins to slide, accelerate to maintain directional stability.

A.  Stop at least a vehicles length before the slope. Select first gear low range with the differential locked.
B.  Engine retardation.
C.  Now unlock differential and change into second or third gear.

**Wading – Fig. ST097**
The maximum advisable wading depth is approximately
0,5 metres (20 in). Before wading make sure that the timing
cover drain plug fitted to the diesel model only; and the
flywheel housing drain plug are in position, see Maintenance
Section, and if the water is deep, slacken off the fan belt.
To prevent saturation of the electrical system and air intake,
avoid excessive engine speed. A low gear with the differential
locked is desirable and sufficient throttle should be
maintained to avoid stalling if the exhaust pipe is under water.

**After being in water**
Make sure that the brakes are dried out immediately so that
they are fully effective when needed again. This can be
accomplished by driving a short distance with the footbrake
applied. Also re-tighten and adjust the fan belt, remove the
flywheel housing drain plug and, on diesel models remove
the timing cover drain plug see Maintenance Section for
these operations.
Do not rely on the handbrake to hold the vehicle once the
transmission brake has been subjected to mud and water;
leave the vehicle parked in gear.

**Driving in soft, dry sand**
When conditions are soft, reduced tyre pressures will
increase the contact area, help improve traction and reduce
the tendency to sink. Select a gear, lock the differential and
stay in it.

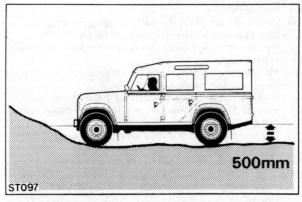

ST097  500mm

Because of the drag of the sand, the instant the clutch is
disengaged the vehicle will stop. If a standing start in sand or
on the side of dunes if necessary, exercise care in applying
the accelerator pedal, as sudden power will induce wheel
spin and cause the vehicle to dig itself into trouble.

**Ice and snow**
Land Rovers are, of course, used extensively in snow and icy
conditions. The driving techniques are generally the same as
driving on mud or wet grass. Select the highest gear possible
with the differential locked and use only sufficient engine
revolutions to just move the vehicle forward without
labouring. Avoid violent movements of the steering wheel
and use the brakes, with care, only if necessary.

## Towing

The weight of the trailer plus load depends upon several factors.

(a) Towing stability.

(b) Weight of the vehicle contents including passengers. When part of the weight is transferable, loading the towing vehicle will generally improve the stability of the combination.

(c) Altitude: Engine performance is progressively reduced above altitudes of 300 m (1,000 feet).

(d) For trailer stability (2 wheel trailers) the load imposed on the vehicle tow bar (nose weight) should be maximum 75 kg (165 lbs).
See recommended maximum laden trailer weights specified in Data Section.

Vehicle tyre pressures must always be set at normal pressures, NOT the 'reduced for comfort' option, irrespective of the load being carried. See 'Tyre pressures' in Data Section.

It is the driver's responsibility to ensure that all regulations with regard to towing are complied with according to the territory in which the vehicle is operated. All relevant information should be obtained from an appropriate motoring organisation.

## Vehicle recovery – towed

If the vehicle should suffer a breakdown or accident damage and it becomes neccessary to make a towed recovery, it is essential to adhere to one of the following procedures depending on the type of tow to be undertaken.

This is because Land Rovers have permanent four-wheel drive and may be fitted with a steering lock.

## Towing the Land Rover (on four-wheels)

(a) Set the main gearbox in neutral.

(b) Set the transfer box in neutral.

(c) Turn the starter/steering lock key to the first position to unlock the steering.

(d) Ensure the differential lock is in the normal 'unlocked' position. If, on V8 models, the differential is 'locked', start the engine to provide vacuum operation to switch to 'unlock'. If the engine is inoperative, remove one of the propeller shafts.

(e) Secure towing attachment to the vehicle.

(f) Release the handbrake.

**NOTE:** Unless the engine is running, brake servo cannot be maintained. This will result in a considerable increase in pedal pressure being required to apply the brakes.

**CAUTION:** Where a front propeller shaft is to be removed check whether the four rear end fixing bolts in the gearbox flange are entered from the gearbox side. In this event they cannot readily be withdrawn. However, since the flange will revolve as soon as the vehicle is towed the four loose bolts MUST be tightly secured with nuts or suitably wired to prevent damage to the gearbox end casing.

**Suspended tow by breakdown vehicle**

Disconnect the propeller shaft from the axle to be trailed; this is necessary as the vehicle has a permanent four-wheel drive. If the front axle is to be trailed it will also be necessary to turn the starter steering lock key to position 1 to unlock the steering.

The steering wheel and/or linkage MUST be secured in a straight ahead position.

The vehicle can then be attached to the breakdown vehicle and raised.

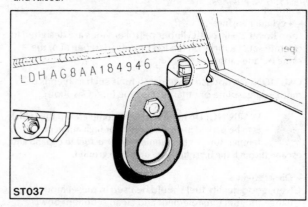

ST037

**Transporting the Land Rover on a trailer – Fig. ST037**

Lashing rings are available on the front and rear chassis members to facilitate the securing of the vehicle to a trailer.

NOTE: Since towing regulations vary from country to country, it is important to refer to the relevant national motoring organisations for the laws relating to towing weights and speed limits.

The following maximum permissible towed weights refer to the design limitations.

| Maximum Permissible Towed Weights | | On-road kg | Off-road kg |
|---|---|---|---|
| Unbraked trailers | | 750 | 500 |
| Trailers with overrun brakes | | 3500 | 1000 |
| 4 wheel trailers with coupled brakes | Diesel models **(except Turbo)** | 3500 | 1000 |
| | * Petrol and Diesel Turbo | 4000 | 1000 |

* NOTE: In order to tow a trailer with a weight in excess of 3,500 Kg, it is necessary to adapt the vehicle to operate a Coupled Brake System, and the VIN plate must be changed to show the increased train weight.

A revised VIN plate may be obtained from Land Rover, which will be issued, subject to satisfactory proof that the vehicle has been fitted with an approved conversion.

## Running-in period

Progressive running-in of a new Land Rover is important and has a direct bearing on reliability and smooth running throughout its life.

The most important point is not to hold the vehicle on large throttle opening for any sustained periods. To start with, the maximum speed should be limited to 65 to 80 km/h (40 to 50 mph) for 4-cylinder models and 80 to 95 km/h (50 to 60 mph) for V8 cylinder models, on a light throttle and this may be progressively increased over the first 2.500 km (1,500 miles).

## Tools

The small tools are carried in a locker, under the seat cushion. On some vehicles, the lifting jack is secured in clips on the seat backrest panel and is accessible with the seat backs lowered.

## Cleaning the vehicle

Use a sponge and plenty of water to clean the exterior.

CAUTION: DO NOT use water to clean the dash panel, as it could enter the fuse box and switches causing damage.

## Fuel recommendations

Recommended fuels for petrol models are specified in the Data section. No advantage will be gained by the use of higher octane fuels.

### – Unleaded petrol

Petrol with minimal quantities of lead is now available in certain territories.

### – V8 engines

Land Rover V8 engines are designed to operate satisfactorily on leaded or lead free petrol of the same octane rating. However, since the most commonly available *lead free petrol* is 91 octane (R.O.N.) it is only suitable *for low compression (8.13:1) V8 engines.*

### – 4 cylinder engines

Land Rover 2.5 litre, 4-cylinder petrol engines are designed to operate satisfactorily on leaded or unleaded petrol of the same octane rating.

CAUTION: Do not use oxygenated fuels such as blends of methanol/gasoline or ethanol/gasoline (e.g. Gasohol).

 WARNING: Do not fill the tank completely if the vehicle is to be parked in direct sunlight or high ambient temperature, as this would cause the fuel to expand and escape through the breather pipe on to the ground.

### – Diesel engines

Clean, good quality fuel should be used in diesel models. Change the fuel filter element and clean sediment bowl regularly.
The fuel filler cap is located:
Side tank: at the front right-hand side of the body.
Rear tank: at the rear right-hand side of the body.

**Basic attention**
In addition to the Workshop Maintenance Schedules shown
in Section 5, the following checks and adjustments should be
carried out by the driver or operator, to ensure that the
vehicle is ready for daily use.
Many of these tasks are described and illustrated in the
following pages.
Recommended lubricants, fluids and quantities are stated in
Section 6.

**Daily or weekly,** depending on operating conditions, and at
least every 500 km (250 miles):-

 Check/top up engine oil.
 Check/top up radiator cooling system.
 Check/top up windscreen washer reservoir.
 Check/top up brake fluid reservoir.
 Check/top up clutch fluid reservoir.
 Check/top up power steering reservoir.
 Check/adjust tyre pressures.
 Check tyres for wear or damage.
 Check operation of handbrake and footbrakes.
 Check operation of all lights and horn.

**Air cleaner**
When the vehicle is used in dusty or field conditions or deep
wading, frequent attention to the air cleaner may be required.
See details in Section 5.

**Spark plugs**
Cleaning and adjustment is detailed in Section 5.

**Camshaft drive belt – 2½ litre diesel engines**
The engine timing gears are driven by a flexible rubber belt
which must be renewed at intervals determined by the
severity of operating conditions.

**In reasonable, temperate climate operation,** renew the belt
every 100,000 km (60,000 miles) or every five years whichever
occurs earlier.

**In adverse operating conditions** such as work in dusty
atmospheres or in high ambient temperatures, renew the belt
every 50,000 km (30,000 miles) or every two and a half years
whichever occurs earlier.

**In severe operating conditions** such as in arid desert and
tropical zones, renew the belt every 40,000 km (24,000 miles)
or every two years whichever occurs earlier.

If the drive belt is not renewed at the correct interval, it could
fail, resulting in serious engine damage.

**Brakes, servo assistance (where fitted)**
Never coast with the engine switched off as the brake servo
will not operate. The brakes will still function but more foot
pressure will be required.

55

## Exterior lamps

Owners are under a legal obligation in many territories to maintain all exterior lights in good working order; this also applies to headlamp beam setting, which should be checked at regular intervals by your Distributor or Dealer.

## Snow chains

Chains may be fitted to provide increased traction during extremely adverse heavy snow conditions.
Never fit chains to one wheel only, always fit snow chains in pairs to the rear axle only, and ensure the gearbox differential control is in the LOCKED position.
Remove the snow chains immediately the road is clear of snow.

## Spare wheel

The spare wheel stowage position varies on different models as follows:
It can be mounted in a well in front of the wheel arch panel or on the rear door on station wagons.
It can also be fitted to the bonnet top panel on all models, using a specially adapted bonnet.

## Lifting jack types

Two different types of lifting jack are described in the following instructions. Refer to the instructions applicable to the jack being used.

## Before jacking the vehicle

It is most important that the jacking procedure, described in this manual, is followed. Wheels should be chocked in all circumstances.

 **WARNING: The handbrake acts on the transmission, not the rear wheels and may not hold the vehicle when jacking unless the following procedure is used. If one front wheel and one rear wheel are raised no vehicle holding or braking effect is possible.**
**Wheels should be chocked in all circumstances.**

1. The jack should be used on level and firm ground.
2. Always engage the differential lock before jacking. The differential lock is only engaged if the warning light is illuminated with the ignition switched on.
3. No person should remain in a vehicle being jacked.
4. Apply the handbrake.
5. Engage first gear in the main gearbox.
6. Engage low gear in the transfer box.

**Bottle jack – Fig. ST081 – suitable for all models**
**To jack up a front wheel:** Jack up the corner of the vehicle by positioning the jack so that when raised, it will engage with the front axle casing immediately below the coil spring where it will be located between the flange at the end of the axle casing and a large bracket to which front suspension members are mounted.

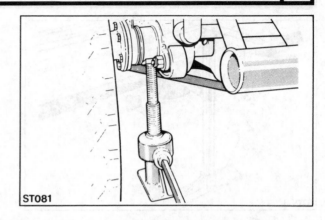
ST081

**To jack up a rear wheel – Fig. ST082:** Jack up the corner of the vehicle by positioning the jack so that when raised, it will engage with the rear axle casing immediately below the coil spring and as close to the shock absorber mounting bracket as possible.

 **WARNING: It is unsafe to work under the vehicle using only the jack to support it. Always use stands or other suitable supports to provide adequate safety.**
Neglect of the jack may lead to difficulty in a roadside emergency. Examine the jack occasionally, clean and grease the thread to prevent the formation of rust.
When the jack is not in use, it should be retained in its stowage position with the clips provided.

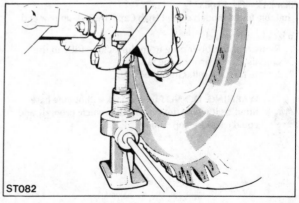
ST082

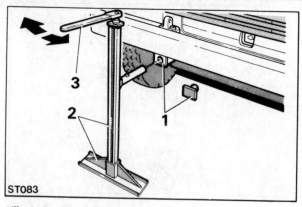

ST083

**Pillar jack – Fig. ST083**
**– suitable for all models except High Capacity Pick-up rear end**

**To jack up any wheel:**

1. Remove the rubber plug from the jacking tube in the chassis at the corner to be raised.
2. Locate the jack pillar into the base.

 **WARNING: DO NOT use the jack without the base fitted, as it would not support the vehicle properly and could cause personal injury.**

3. Fit the handle to the jack and adjust the height of the jacking peg until it can be lifted into the jacking tube. Note that the jack handle has a ratchet, use one side to raise the jack, turn the handle over to lower the jack.
4. Ensure that the jacking peg is pushed as far as possible into the jacking tube and that the pillar is upright, then operate the jack handle to raise the vehicle.

**Wheel changing**

1. Using the wheel nut wrench supplied in the tool kit, initially slacken the nuts on the wheel to be removed before jacking the vehicle.
2. Jack up the corner of the vehicle.
3. When the wheel is clear of the ground, remove wheel nuts and lift off wheel.
4. If available, place a drop of oil or grease on the wheel studs to assist in replacement.
   Fit spare wheel; tighten the nuts as much as possible.
5. Lower the vehicle to the ground and finally tighten the nuts to the following torque:
   10,4 to 11,7 kgf m (75 to 85 lbf ft).

Remember to disengage the differential lock after road wheel has been replaced.

**Road wheel nuts**
Check road wheel nuts for tightness, torque 10,4 to 11,7 kgf m (75 to 85 lbf ft). **DO NOT overtighten.**
When using the wheelbrace from the vehicle tool kit apply hand pressure only. DO NOT use foot pressure or extension tubes as this could overstress the wheel studs.

**Engine oil level – 4-cylinder engines – Fig. ST265**
Check daily or weekly depending on operating conditions
and at least every 500 km (250 miles). The oil level should not
be allowed to fall below the 'L' (low) notch on the dipstick (2)
located on the left-hand side of the engine.
Whenever possible, the oil level should be checked with the
engine hot, as follows:
Stand the vehicle on level ground and wait at least five
minutes, after the engine has stopped, for the oil to drain
back into the engine sump.
Withdraw the dipstick at the left-hand side of the engine,
wipe it clean, re-insert it to its full depth and remove a second
time to take a reading.
If the oil level is in the upper half of the distance between the
'L' (lower notch) and the 'H' (upper notch) on the dipstick,
add no oil. If it is in the lower half above the 'L' notch, add
one litre of oil only. The oil filler is located on the rocker
cover and is fitted with a push on breather cap (3) which is
easily removed for application of the correct grade of oil as
necessary.
If the oil level is below the 'L' notch, add two litres of oil and
re-check the level after five minutes.

**If the engine is cold:**
**DO NOT** start the engine. Ensure that the vehicle is standing
on level ground and proceed as above.
If it is necessary to recheck oil, or if the engine has been
started without being thoroughly warmed up, wait at least 30
minutes to confirm oil level.

**DANGER**
**OIL LEVEL MUST NEVER BE ABOVE THE 'H' NOTCH AS
ENGINE DAMAGE MAY BE CAUSED.**

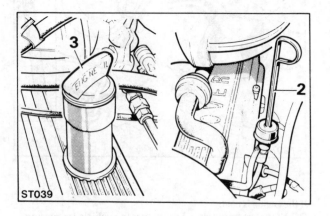

ST039

**Engine oil level – V8 cylinder petrol engines – Fig. ST039**
Check daily or weekly, depending on operating conditions
and at least every 500 km (250 miles).
Whenever possible, the oil level should be checked with the
**engine hot,** as follows:

1. Stand the vehicle on level ground and wait at least five
   minutes, after the engine has stopped, for the oil to drain
   back into the engine sump.
2. Withdraw the dipstick at the left-hand side of the engine,
   wipe it clean, re-insert it to its full depth and remove a
   second time to take a reading. The oil level should not be
   allowed to fall below the 'LOW' mark.
3. Add the correct grade of oil, as necessary, through the
   screw-on filler cap marked 'ENGINE OIL' on the
   right-hand front rocker cover. **Never fill above the 'HIGH'
   mark.**

**If the engine is cold:**
**DO NOT** start the engine.

1. Stand the vehicle on level ground.
2. Withdraw the dipstick at the left-hand side of the engine,
   wipe it clean, re-insert it to its full depth and remove a
   second time to take a reading. The oil level should not be
   allowed to fall below the 'LOW' mark.
3. Add the correct grade of oil, as necessary, through the
   screw-on filler cap marked 'ENGINE OIL' on the
   right-hand front rocker cover. **Never fill above the 'HIGH'
   mark.**

**Engine coolant**
The coolant level should be checked daily or weekly depending upon the operating conditions.

**Diesel models**
Never run the engine without coolant, not even for a very brief period, otherwise the injectors may be seriously damaged. This is due to the very high rate of heat transfer in the region of the injector nozzles.

**Radiator coolant level – Fig. ST040**
To prevent frost damage or corrosion of engine parts it is imperative that the cooling system is filled with a solution of clean water and the correct type of anti-freeze, winter and summer.

In warm climates where frost precautions are not necessary. A solution of clean water and a corrosion inhibitor (Marstons SQ 36) should be used, this is very important on V8 models, because of the aluminium alloy engine.

NEVER use salt water with anti-freeze or an inhibitor, otherwise corrision will occur. In certain territories where the only available water supply may have some salt content, use only clean rainwater or distilled water.

1. The expansion tank filler cap is under the bonnet.
2. With a cold engine, the expansion tank should be approximately half full.

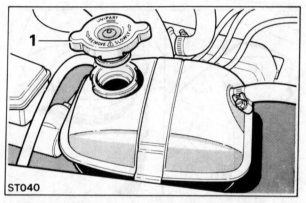

ST040

 **WARNING: Do not remove the filler cap when the engine is hot because the cooling system is pressurised and personal scalding could result.**

3. When removing the filler cap, first turn it anti-clockwise a quarter of a turn and allow all pressure to escape, before turning further in the same direction to lift it off.
4. When replacing the filler cap it is important that it is tightened down fully, not just to the first stop. Failure to tighten the filler cap properly may result in water loss, with possible damage to the engine through overheating.

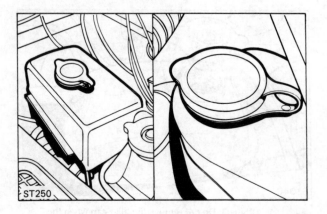

ST250

**Windscreen and rear door washer reservoirs – Fig. ST250**

The windscreen washer reservoir (illustrated), is located in the engine compartment. If a rear screen washer is fitted, the reservoir has a large capacity and is fitted with two pumps, one for the front windscreen and one for the rear. If headlamp washers (option) are also fitted, an additional separate reservoir is also fitted.

1. Remove reservoir cap.
2. Top-up reservoir to within approximately 25 mm (1 in) below bottom of filler neck.
3. Use a screen washer solvent in the container; this will assist in removing mud, flies and road film.
4. In cold weather, to prevent freezing of the water, add methylated spirits.

## Carburetters

Carburetter mixture ratio and idle speed settings are pre-set at manufacture and must not be interfered with. Under normal circumstances they do not require attention except at major engine overhaul.

However, should it become necessary to check any aspect of carburetter adjustment the work must be carried out by a qualified Land Rover Distributor or Dealer, who has the specialised equipment needed to carry out adjustments to the close limits necessary to ensure that the engine conforms to the legal requirements in respect of exhaust emission.

**European Countries — Under no circumstances must the mixture setting be disturbed, as this would almost certainly result in the vehicle failing to meet with legal requirements in respect of air pollution.**

ST262

### Carburetter hydraulic damper – V8 cylinder models – Fig. ST262

1. Unscrew the cap on top of the suction chamber, withdraw cap and plunger. Top up with clean engine oil to bring the level to the top of the hollow piston rod. Screw the cap firmly into the carburetter.

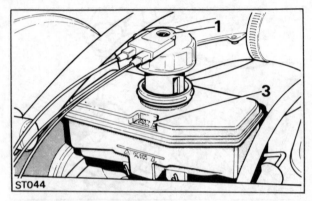

ST044

**Brake fluid reservoir – Fig. ST044**
The tandem brake fluid reservoir is integral with the servo unit and master cylinder. Check the fluid level as follows:

1. Hold the centre terminal block stationary and unscrew the reservoir filler cap.
2. Check the fluid level in the reservoir. The level is indicated on the translucent reservoir body.
3. Top-up if necessary with fluid specified in the Data section to the MAX mark.
4. Replace the filler cap.

If significant topping-up is required, check master cylinder, wheel cylinders and brake pipes for leakage; any leakage must be rectified immediately.

**CAUTION:** When topping-up the reservoir, care should be taken to ensure that brake fluid does not come into contact with any paintwork on the vehicle.

Where a vehicle is operated in extremely dusty conditions, consult your Land Rover Distributor or Dealer for advice on servo air filter change intervals. The filter is situated on the brake pedal side of the servo unit.

**Low fluid level/Brake circuit warning light**

 **WARNING: As this test requires the release of the handbrake ensure the vehicle is on level ground.**

Normally the warning light remains off, however to check that the circuit is operative, switch on the ignition and release the handbrake. Press the flexible contact located in the filler cap centre, the RED warning light on the instrument panel should illuminate; if it does not energise, and the bulb has not failed, consult your Land Rover Distributor or Dealer immediately.

## Clutch fluid reservoir – Fig. ST043

Check the fluid level in the reservoir, mounted on the
bulkhead adjacent to the brake servo.

1. Remove the cap; top-up if necessary to bottom of filler
   neck. Use the correct fluid specified in Data section.
2. If significant topping-up is required, check for leaks at
   master cylinder, slave cylinder and connecting pipes.

**CAUTION:** When topping-up the reservoir, care should be
taken to ensure that brake fluid does not come into contact
with any paintwork on the vehicle.

ST043

## Check/top-up power steering reservoir – Fig. RR1240

The power steering units are lubricated by the operating
fluid. The only lubrication attention required is to check the
reservoir level as follows:

Unscrew the fluid reservoir cap (1) which is fitted with a
dipstick.

Check that the fluid is up to the high mark on the dipstick.

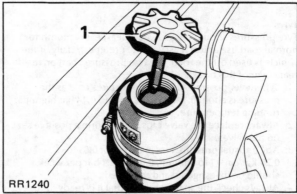

RR1240

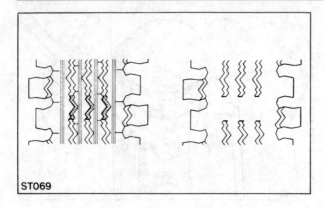

ST069

**Tyres**

Tyre pressures should be checked at least every month for normal road use and at least weekly, preferably daily, if the vehicle is used off the road. See tyre pressure chart on inside rear cover of this book.

1. Whenever possible check with the tyres cold as the pressure is about 0,1 kgf/cm² (2 lbf/in²) 0,14 bar higher at running temperature.
2. Always replace the valve caps as they form a positive seal on the valves.
3. Any unusual pressure loss in excess of 0,05 to 0,20 kgf/cm² (1 to 3 lbf/in²) 0,07 to 0,21 bar per week should be investigated and corrected.
4. Always check the spare wheel so that it is ready for use at any time.

5. At the same time remove embedded flints etc. from the tyre treads with the aid of a penknife or similar tool and check that the tyres have no breaks in the fabric or cuts to sidewalls etc. Clean off any oil or grease on the tyres using white spirit sparingly.
6. Check that there are no lumps or bulges in the tyres or exposure of the ply or cord structure.
7. 'Butyl' synthetic innertubes are fitted and all repairs must be vulcanised.
8. Maximum tyre life and performance will only be obtained if the tyres are maintained at the correct pressure.

It is illegal in the UK and many other countries to continue to use tyres with excessively worn tread. Tyre wear should be checked at every maintenance inspection.

**Check tyres for tread depth and visually for external cuts in the fabric, exposure of ply or cord structure – Fig. ST069**
Most tyres fitted to Land Rovers as original equipment include wear indicators in their tread pattern. When the tread has worn to a remaining depth of 1,6 mm ($\frac{1}{16}$ in) the indicators appear at the surface as bars which connect the tread pattern across the full width of the tyre, as in the tyre section illustrated. When the indicators appear in two or more adjacent grooves, at three locations around the tyre, a new tyre should be fitted. If the tyres do not have wear indicators, the tread should be measured at every maintenance inspection and when the tread has worn to a remaining depth of 1,6 mm ($\frac{1}{16}$ in), new tyres should be fitted. Do not continue to use tyres that have worn to the recommended limit or the safety of the vehicle could be affected and legal regulations governing tread depth may be broken.

**WARNING**

1. Do not let the engine run without battery connected.
2. Do not use a high-speed battery charger as a starting aid.
3. When using a high-speed charger to charge the battery, the battery must be disconnected from the rest of the vehicle's electrical system.
4. When installing, ensure that the battery is connected with the correct polarity.
5. No larger battery than 12v must be used.
6. DO NOT use steam to clean the engine compartment.
7. The battery MUST be disconnected before carrying out any electrical welding on the vehicle.
8. IF A REPLACEMENT BATTERY IS FITTED TO THE VEHICLE, IT SHOULD BE THE SAME TYPE AS THE ORIGINAL BATTERY. ALTERNATIVE BATTERIES MAY VARY IN SIZE AND TERMINAL POSITIONS; AND THIS COULD BE A POSSIBLE FIRE HAZARD IF THE TERMINALS OR LEADS COME INTO CONTACT WITH THE BATTERY CLAMP ASSEMBLY. WHEN FITTING A NEW BATTERY ENSURE THAT THE TERMINALS AND LEADS ARE WELL CLEAR OF THE BATTERY CLAMP ASSEMBLY.

ST049

**Battery electrolyte**

A low maintenance battery is installed in the vehicle underneath the left-hand front seat. The battery compartment (Fig. ST049) is accessible by pulling up the front of the seat to release it from retaining clips and drawing it forward. This will reveal the compartment cover which can be removed after release of a catch on the front edge. Dependent upon climate conditions the electrolyte levels should be checked as follows:
Temperature climates every 3 years. Hot climates every year.
The exterior of the battery should be occasionally wiped clean to remove any dirt or grease. Periodically remove the battery terminals to clean and coat with petroleum jelly. To check if maintenance is required, gently prise off the vent covers and inspect the electrolyte level of the centre cell. This should be no lower than 1 mm (0.04 in) above the top of the plates. If necessary, top up (with distilled water only) to a maximum of 3 mm (0.12 in) above the plates.

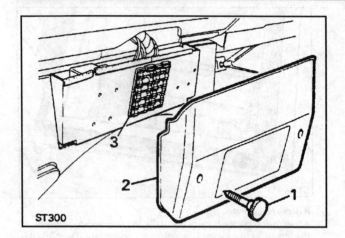

ST300

## TO REPLACE A FUSE
Unscrew the two knobs (1).
Pull off the fuse box cover (2).
Replace fuse (3) as required.
Refit the fuse box cover.

**FUSE BOX - Fig. ST300**
The fuse box is located in the centre of the dash in front of the main gear change lever and
contains twelve fuses of the following values:

| | | |
|---|---|---|
| 1 AMP | or | 1 HOLD |
| 5 AMP | or | 5 HOLD |
| 8 AMP | or | 7.5 HOLD |
| 10 AMP | or | 10 HOLD |
| 12 AMP | or | 12.5 HOLD |
| 17 AMP | or | 17.5 HOLD |

Only glass type fuses of the correct rating should be used as replacements. The location of the
fuses and the items they protect are shown in the chart attached to the fuse box cover.

**NOTE: Early vehicles may be fitted with a 2.5 amp fuse and this may be specified on the
chart inside the fuse box cover, but it should always be replaced with a 5 amp fuse.**

**HEADLAMPS - Fig. ST301**
To replace light unit or bulb:

**CAUTION:** DO NOT touch the glass on "Halogen" bulbs with the fingers, as this could damage the bulb. If contact is accidently made, wipe the glass gently with methylated spirits.
Remove the screw and lens (1) from the side and flasher lamps.
Remove the screws and pull the lamp back plates (2) forward, as far as the leads allow.
Remove the screws retaining the plastic finisher (3) for the headlamp and move the finisher aside.
Remove the three recessed head screws (4) retaining the headlamp rim.
Remove the rim (5).
Lift out light unit (6) and pull off electrical connector.
Remove from connector the rubber grommet (7).
Bulb or light unit as applicable can now be replaced.
Refit rim and headlamp finisher.

**CAUTION:** Fitting headlamp bulbs or light units with a higher watt value than the Specification in the Data Section, will result in damage to the 'Dim Dip' unit (if fitted), wiring and switches.

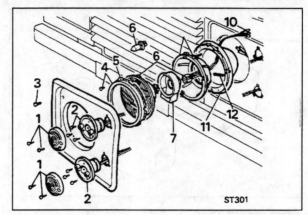

ST301

**Side, tail, stop and flasher lamp – Fig. ST073**
To replace a bulb:
1. Remove the retaining screws and withdraw the lens.
2. Renew the bulb.
3. Replace the lens and retaining screws.

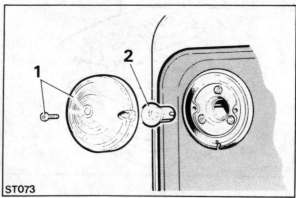

ST073

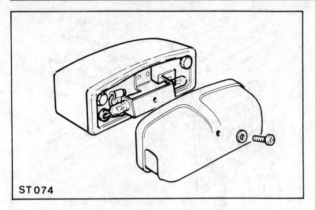

ST 074

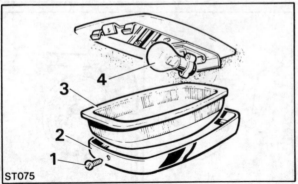

ST075

**Rear number plate lamp (where applicable) – Fig. ST074**
To replace the bulbs:
1. Slacken the securing screw.
2. Remove cover.
3. Bulbs are then accessible in the lamp body.
4. Replace bulbs and refit cover.

**Interior light (where applicable) – Fig. ST075**
To replace the bulb:
1. Remove screw retaining rim and cover.
2. Remove the rim.
3. Remove retaining cover.
4. Replace bulb.
5. Refit cover and trim.

**Warning lights – Fig. ST079**

To replace a bulb:

1. Disconnect the battery.
2. Remove two screws and withdraw the warning light module from the front of the instrument panel.
3. Pull off plug connector to give access to warning light bulbs.
4. Twist the bulb holder and pull it from its socket.
5. Pull the bulb from the holder.
6. Fit a new bulb and refit holder and plug connector.
7. Refit module and reconnect battery.

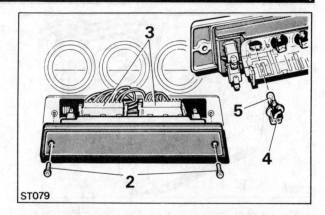

**Instrument illumination – Fig. ST078**

To replace a bulb:

1. Disconnect the battery.
2. Remove four screws retaining the instrument panel.
3. The instrument panel can now be eased forward for access to the bulbs. If necessary, disconnect the drive cable from the back of the speedometer.
4. Twist the bulb holder and pull it from its socket.
5. Pull the bulb from the holder.
6. Fit new bulb and refit holder.
7. Replace instrument panel.
8. Reconnect the battery.

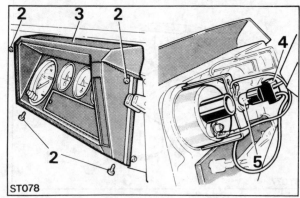

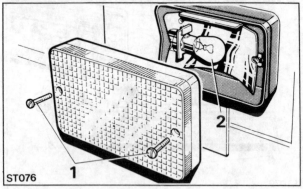

ST076

**Reversing and rear fog guard lamp – Fig. ST076**

To replace the bulb:
1. Remove the retaining screws and withdraw the lens.
2. Renew the bulb.
3. Replace the lens and retaining screws.

**Direction indicator side repeater lamps, on front wings (when fitted) – Fig. ST077**

1. Remove screw retaining the front of the lens.
2. Slide the lens forward to disengage it from the lamp base projection.
3. Remove the lens (and gasket if loose).
4. Renew the bulb.
5. Reset the gasket and replace the lens and fix by screw.

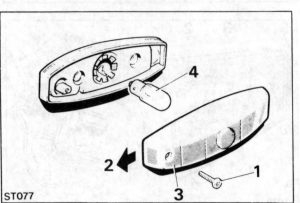

ST077

**Windscreen and rear door wiper blades**
Check, and if necessary, renew wiper blades.

### Windscreen – Fig. LR2076
1. Lift the wiper arm away from the windscreen.
2. Squeeze the spring clip and push the wiper blade towards the windscreen and unhook it from the wiper arm.
3. To fit a new blade, push it over the arm and hook the arm into the swivel bracket ensuring that the retaining clip is engaged.

### Rear door – Fig. ST080
1. Lift the wiper arm forwards, away from the rear door.
2. Twist the wiper fixing bracket in the direction arrowed and disengage it from the wiper arm.
3. To fit a new blade locate its fixing bracket over the end of the wiper arm and push on until the retaining dowel is engaged.

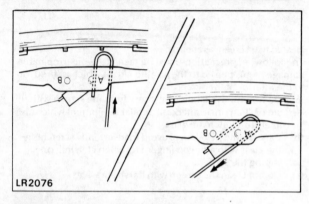

LR2076

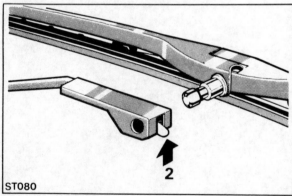

ST080

**Heated rear screen, as applicable**

The following precautions must be taken to avoid irreparable damage being caused to the printed circuit which is 'fired' on to the interior of the screen.

(a)  Do not remove labels or stickers from the screen with the aid of sharp instruments or similar equipment which are likely to scratch the glass.

(b)  Care should be taken to avoid inadvertently scratching the glass with a ringed finger etc. when cleaning or wiping the screen.

(c)  Do not clean the screen with harsh abrasives.

These engine fault diagnosis pages have been compiled for your general assistance in an emergency.
Checking of any part of the electronic ignition system must be referred to your Land Rover Dealer or Distributor.

 **WARNING: The electronic ignition system involves very high voltages. Inexperienced personnel and wearers of medical pacemaker devices should not be allowed near any part of the high-tension circuit.**

| Sympton | Fault | Remedy |
|---|---|---|
| Starter will not turn engine (headlights dim) | Battery low in charge, often causing the solenoid to chatter | Charge battery and check charging system |
| | Defective battery | Renew battery |
| | Corroded battery cables or loose connections | Clean battery connections or renew battery cables. Tighten battery and starter-motor connections. |
| (Headlights bright) | Starter jammed. | Free starter |
| | Defective starter solenoid | Renew |
| | Defective starter | Renew or overhaul |
| | Defective starter switch | Renew |
| Engine turns slowly but will not start | Battery low in charge | Charge battery and check charging system |
| | Defective battery | Renew |
| | Corroded battery cables or loose connections | Clean and secure battery connections |
| | Poor engine-to-chassis earth strap connections | Clean and secure connection |
| | Defective starter | Renew or overhaul |
| Engine turns normally but will not fire | Ignition fault | Check for spark at plug lead. Note above warning |
| | Where no spark is observed at plug lead | Consult your Dealer or Distributor |
| | Where spark is observed at plug lead | With engine cold, check mixture control operation. If necessary, a drop of oil on the butterfly spindle may help. Loosen petrol-pipe union at carburetter. Switch on ignition for electric pump. Check if petrol is being delivered. |
| | No fuel to carburetter | Remove petrol-tank cap and check for fuel (fuel gauge may be inaccurate) |
| Engine backfires violently, kicks back or bangs through carburetter | Ignition timing faulty | Consult your Dealer |
| | Damp distributor cap and leads | Dry thoroughly and check firing order |
| Engine fires, but fails to keep running | Ignition or fuel fault | Refer to order of checks for 'Engine turns normally but will not fire', with special emphasis on mixture control, plug condition and continuous HT spark at plug lead |

| Symptom | Fault | Remedy |
|---|---|---|
| Engine stalls when idling (engine cold) | Mixture control throttle-stop requires adjustment | Adjust |
| | Mixture control not operating correctly | Check mixture control operation |
| Engine stalls when idling (engine hot) | Engine idle speed too low | Adjust idle speed |
| | Mixture control stuck in operation | Check mixture control operation at carburetter |
| | Carburetter flooding | Adjust fuel level or float setting to specification. Clean needle valve |
| | Intake vacuum leak | Check manifold, carburetter mounting, any connections to manifold and vacuum advance. Also check butterfly spindle and bosses; if worn, seek advice. |
| Engine has rough idle | Fouled or improperly gapped spark-plugs | Clean and adjust plug gaps or renew plugs |
| | Incorrect ignition timing | Consult your Dealer |
| | Intake vacuum leak | Check manifold, carburetter mounting, any connections to manifold and vacuum advance |
| Engine stalls on acceleration | Mixture control not functioning properly, or improperly adjusted | Check mixture control operation at carburetter with engine cold |
| | Insuffient fuel supply to carburetter | Clean needle valve and jets. Check float level |
| | Air-cleaner element dirty | Clean or renew filter element. Conform to recommended maintenance schedule |
| | Carburetter piston seized – (V8 SU carburetter) | Polish piston and cylinder with dry or petrol-damp rag. Check that correct oil is used in dash-pot and top up to the required level |
| Engine has poor acceleration | Incorrect ignition timing | Adjust |
| | Intake vacuum leak. | Tighten or renew faulty gaskets |
| | Insufficient fuel supply | Clean needle valve and jets. Check fuel |
| | Accelerator linkage out of adjustment | Check that full throttle on the pedal is also full throttle at carburetter. Adjust as necessary |

| Sympton | Fault | Remedy |
| --- | --- | --- |
| Failure to start | Fuel tank empty | Fill tank and bleed air |
| — Fuel | Obstructed fuel lines | Purge fuel lines |
| | Fuel filter clogged | Renew filter and bleed system of air |
| | Fuel line leakage | Check all fuel line connections for tightness |
| | Air in fuel system | Bleed fuel system |
| | Faulty fuel lift pump | Renew or overhaul pump |
| | Clogged air cleaner | Clean or renew filter |
| — Electrical | Discharged battery | Check electrolyte and recharge |
| | Loose or corroded cable connections | Clean and tighten battery terminals |
| | Faulty starter switch | Renew starter switch |
| | Faulty starter | Renew/repair starter |
| | Faulty heater plugs | Renew heater plugs |
| | Faulty injection pump solenoid valve | Check electrical supply. Renew solenoid valve |
| Starts and stops | Obstructed fuel lines | Purge fuel lines |
| | Fuel filter clogged | Renew filter and bleed system of air |
| | Air in fuel system | Bleed fuel system |
| | Low idle speed | Check and adjust minimum engine rev/min |
| | Faulty fuel lift pump | Renew or overhaul pump |
| | Clogged fuel tank breather | Clear fuel tank breather |
| Poor engine performance | Insufficient fuel supply to injectors | Check for fuel leaks, air in system, clogged fuel filter, wrong type or contaminated fuel |
| | Valve clearance out of adjustment | Adjust valve clearances |
| | Faulty injection pump timing | Check/adjust pump timing or consult Dealer |
| | Injector fault | Renew or overhaul injectors |
| Exhaust smokes badly | Injector fault | Test/adjust injectors |
| | Faulty injection pump timing | Check/adjust pump timing or consult Dealer |
| | Clogged air cleaners | Clean/renew filter |
| | Oil sump overfilled | Drain to proper level |
| Engine overheats | Low coolant level | Fill the cooling system to correct level |
| | Faulty water pump | Renew pump |
| | Water leaks | Check hoses, fittings, plugs and radiator |
| | Faulty temperature gauge | Renew gauge |
| | Low oil level | Fill to proper level with recommended make and grade of oil |
| | Clogged cooling system | Drain and flush cooling system |
| | Incorrect injection pump timing | Adjust timing or consult Dealer |

RR064

### The new vehicle

With every new Land Rover special literature is provided which should be read by all owners and drivers to help obtain the best operating results.

The literature consists of the following:

1. Handbook: This book, which you are now reading, gives general information about the Land Rover, also incorporates notes on service, the Vehicle Service Statement and full information on how to carry out the necessary day-to-day running maintenance.

2. Service Record which gives details of the maintenance required and includes spaces for the Distributor or Dealer to sign and stamp to certify that the work has been carried out at the appropriate intervals.

The operations carried out by your Distributor or Dealer will be in accordance with current recommendations and may be subject to revision from time to time.

Upon receiving the new Land Rover the owner should immediately:

3. Examine the Handbook for advice on new features and as an aid to getting the best out of the vehicle.

4. Arrange with a Land Rover Distributor or Dealer to carry out regular maintenance attention.

### Vehicle Service Statement (Warranty)

Land Rover issue under the heading of Vehicle Service Statement an undertaking regarding its Service Policy.

Home market: The Vehicle Service statement is supplied in the Literature Pack.

Export markets: The Warranty, Vehicle Service Statement, should be obtained from the Distributor or Dealer at the time of purchase.

The following notes are given for guidance in the event of a claim being put forward:

1. The Land Rover or the part in respect of which a claim is made must be taken immediately to a Land Rover Distributor or Dealer. This should, wherever possible, be the Distributor or Dealer responsible for the sale of the vehicle to the owner.

2. The Distributor or Dealer will examine the parts or Land Rover and will without charge advise on the action to be taken in respect of the claim. It will be noted that the Company must reserve the right to examine any alleged defective parts or material should they think fit prior to the settlement of any claim.

3. It must be understood that the factors of wear and tear and any possible lack of maintenance or unapproved alteration will be taken into consideration in respect of any claim submitted.

4. It will be noted that tyres and glass are expressly excluded. The manufacturers of those tyres which the Company fits as standard to its vehicle will always be prepared to consider any genuine claim.

5. It is recommended that owners should arrange with their Insurance Company to provide separate cover for the glass at the small extra cost involved.

## Spare parts and accessories

When new parts or accessories are required, obtain Genuine Land Rover parts, or parts supplied through sources approved by the Company.

Land Rover Distributors and Dealers are obligated to supply only such parts.

Through other sources parts are often sold as being suitable for Land Rovers but frequently these are not made to the same standard or specification as the Company parts and are therefore less likely to give the requisite performance.

**Genuine Land Rover parts and accessories are designed and tested for your vehicle and have the full backing of the Land Rover Vehicle Service Statement. ONLY WHEN GENUINE LAND ROVER PARTS ARE USED CAN RESPONSIBILTY BE CONSIDERED UNDER THE TERMS OF THE STATEMENT.**

Safety features embodied in the vehicle may be impaired if other than genuine parts are fitted. In certain territories, legislation prohibits the fitting of parts not to the vehicle manufacturer's specification. Owners purchasing accessories while travelling abroad should ensure that the accessory and its fitted location on the vehicle conform to mandatory requirements existing in their country of origin.

## Maintenance attention

Efficient maintenance is one of the main factors in ensuring continuing reliability and efficiency. For this reason detailed schedules have been prepared so that at the appropriate mileages or times owners may know what is required.

The **Pre-delivery Inspection** is a very important first step in the work of preventive maintenance. The Dealer responsible for the sale of the new Land Rover will have completed the work involved.

There is provision in the Service Record for certification that this work has been carried out.

**Normal day-to-day attention** required is described in Section 4 of this handbook.

The **first Service Inspection** should be carried out by the Dealer responsible for the sale of the Land Rover to the owner at or about 1500 km (1000 miles). A charge is made only for the lubricants, etc. used in carrying out the service.

Where for any reason it is not convenient for this first service to be carried out by the Dealer responsible for the sale, it can, by prior arrangement, be carried out by any other Land Rover Distributor or Dealer.

## Workshop maintenance schedules

The following maintenance should be carried out by trained personnel in a fully equipped workshop. If the vehicle is operating in a remote area where workshop facilities are not available, maintenance and repair work should be carried out by experienced mechanics in safe conditions.

Maintenance should normally be carried out at 10,000 km (6000 mile) intervals or six months, whichever is first, as described in the following schedules which extend up to 100,000 km (60,000 miles) after which they can be repeated. In severe conditions, such as deep mud or sand, or a very dusty atmosphere, the intervals should be reduced to monthly, weekly or even daily for some items. Ask your Land Rover Dealer for advice.

**At 10.000 km (6,000 miles) or six months, whichever is first.**

Lubricate all locks (not steering lock), hinges and doors — check mechanisms.
Remove road wheels.
Remove road wheel brake drums, wash out dust, inspect shoes for wear and drums for condition.
Inspect wheel cylinders for fluid leaks.
Inspect brake pads for wear, calipers for leaks, and discs for condition.
Refit road wheel brake drums.
Adjust road wheel brakes.
Adjust handbrake if required.
Refit road wheels to original position.
Renew engine oil.
Renew engine oil filter.
Check/top-up gearbox oil.
Check/top-up transfer box oil.
Check/top-up front axle oil.
Check/top-up swivel pin housing oil.
Check/top-up rear axle oil.
Lubricate rear suspension upper link ball joint.
Lubricate propeller shaft universal joints.
Lubricate handbrake mechanical linkage.
Check for oil/fluid leaks from steering and suspension systems.
Check/adjust valve clearance (Turbo-charged diesel).
Clean/adjust spark plugs.
Check crankcase breathing system for leaks, hoses for security and condition.
Top-up carburetter piston dampers.
Check/top-up fluid in power steering reservoir.

Check/top-up clutch fluid reservoir.
Check/top-up brake fluid reservoir.
Check/top-up windscreen and rear washer reservoir.
Check power steering system for leaks, hydraulic pipes and unions for chafing and corrosion.
Check condition of driving belts — adjust if required (not camshaft drive belt — diesel).
Lubricate distributor (not V8).
Check dwell angle — adjust as necessary (not V8).
Check/adjust ignition timing.

**NOTE:** It is important that the ignition timing, dwell angle and carburetter adjustments are set in accordance with the vehicle engine specification and fuel octane rating. Refer to the relevant repair operation manual for details.

Check/adjust engine idle speed and carburetter mixture settings with engine at normal running temperature.
Check operation of air intake temperature control system (V8).
Check/tighten inlet manifold and exhaust manifold bolts (2.5 litre Turbo-Diesel only)
Carry out road or roller test.

 **WARNING: Two wheel roller test must be restricted to 5 km/hour (3 miles/hour). DO NOT engage the differential lock or the vehicle will drive off the roller test rig because the Land Rover is in permanent four wheel drive.**

Check:
For excessive engine noise.
Clutch for slipping/judder/spinning.
Gear selection/noise — high and low range.
Steering for noise/abnormal effort required.
All instruments, pressure, fuel and temperature gauges, warning indicators.
Heated rear screen.
Shock absorbers (irregularities in ride).
Footbrake, on emergency stop, pulling to one side, binding, pedal effort.
Handbrake efficiency.
Road wheel balance.
Transmission for vibrations.
For body noises (squeaks and rattles).
Fuel governor cut-off point.
For excessive exhaust smoke.
Engine idle speed.
Endorse service record.
Report any additional work required.

**At 20.000 km (12,000 miles) or twelve months, whichever is first.**

Check condition and security of seats, seat belt mountings, seat belts and buckles.
Check operation of all lamps.
Check operation of horns.
Check operation of warning indicators.
Check operation of windscreen and rear wipers and washers.
Check condition of wiper blades.
Check security and operation of handbrake.
Check rear view mirror(s) for security, cracks and crazing.
Check operation of all doors, bonnet and tailgate locks.
Check operation of window controls.
Lubricate all locks (not steering lock), hinges and door — check mechanisms.
Lubricate accelerator control linkage and pedal pivot.
Check/adjust tyre pressures including spare.
Check/adjust headlamp alignment.
Check front wheel alignment.
Remove battery connections, clean and grease (refit).
Remove road wheels.
Check tyres comply with Manufacturer's specification.
Check tyres visually for cuts, lumps, bulges, uneven wear and tread depth.
Remove road wheel brake drums, wash out dust, inspect shoes for wear and drums for condition.
Inspect wheel cylinders for fluid leaks.

Inspect brake pads for wear, calipers for leaks, and discs for condition.
Refit road wheel brake drums.
Adjust road wheel brakes.
Adjust handbrake if required.
Refit road wheels to original position.
Renew engine oil.
Renew engine oil filter.
Check/top-up gearbox oil.
Check/top-up transfer box oil.
Check/top-up front axle oil.
Check/top-up swivel pin housing oil.
Check/top-up rear axle oil.
Lubricate rear suspension upper link ball joint.
Lubricate propeller shaft universal joints.
Lubricate handbrake mechanical linkage.
Check visually brake, fuel, clutch pipes/unions for chafing, leaks and corrosion.
Check exhaust system for leakage and security.
Check for oil leaks from engine and transmission.
Check for oil/fluid leaks for steering and suspension systems.
Check axle breather pipes, ensure they are not blocked, pinched or split.
Check security and condition of suspension fixings.
Check condition and security of steering unit, joints and gaiters.
Check tightness of propeller shaft coupling bolts.
Clean fuel sedimenter (diesel only).

Renew fuel filter element (petrol).
Drain flywheel housing if drain plug is fitted for wading (refit).
Clean camshaft drive belt housing filter (diesel).
Check condition of heater plug wiring for fraying, chafing, and deterioration (diesel only).
Renew fuel filter element (diesel).
Check/adjust valve clearance (all models except V8 and Turbo-Diesel).
Renew spark plugs.
Renew air cleaner elements.
Check air cleaner dump valve, clean or renew.
Clean engine breather filter (all models except V8).
Renew engine flame trap(s) (V8).
Check brake servo hose for security and condition.
Check V8 air injection/pulsair system hoses/pipes for security and condition.
Check operation of pulsair check valves.
Check crankcase breathing system tor leaks, hoses for security and condition.
Top-up carburetter piston dampers.
Check/top-up cooling system.
Check/top-up fluid in power steering reservoir.
Check/top-up steering box (manual steering).

*(continued)*

**20.000 km (12,000 miles) or twelve months (continued)**

Check/adjust steering box.
Check/top-up clutch fluid reservoir.
Check/top-up brake fluid reservoir.
Check/top-up windscreen and rear washer reservoir.
Check cooling and heater system for leaks, hoses for security and condition.
Check power steering system for leaks, hydraulic pipes and unions for chafing and corrosion.
Check condition of driving belts — adjust if required (not camshaft drive belt-diesel).
Check ignition wiring and HT leads for fraying, chafing and deterioration.
Clean distributor cap, check for cracks and tracking.
Clean/adjust distributor points (not V8).
Lubricate distributor (not V8).
Check voltage drop between coil CB and earth.
Check dwell angle — adjust as necessary (not V8).
Check/adjust ignition timing.

**NOTE:** It is important that the ignition timing, dwell angle and carburetter adjustments are set in accordance with the vehicle engine specification and fuel octane rating. Refer to the relevant repair operation manual for details.

Check throttle operation.

Check/adjust engine idle speed and carburetter mixture settings with engine at normal running temperature.
Check operation of air intake temperature control system (V8).
Check/tighten inlet manifold and exhaust manifold bolts (2.5 litre Turbo-Diesel only).
Carry out road or roller test:

 **WARNING: Two wheel roller tests must be restricted to 5 km/hour (3 miles/hour). DO NOT engage the differential lock or the vehicle will drive off the roller test rig because the Land Rover is in permanent four wheel drive.**

Check:
For excessive engine noise.
Clutch for slipping/judder/spinning.
Gear selection/noise — high and low range.
Steering for noise/abnormal effort required.
All instruments, pressure, fuel and temperature gauges, warning indicators.
Heater and air conditioning systems.
Heated rear screen.
Shock absorbers (irregularities in ride).
Foot brake, on emergency stop, pulling to one side, binding, pedal effort.
Handbrake efficiency.
Operation of inertia seat belts.
Road wheel balance.

Transmission for vibration.
For body noises (squeaks and rattles).
Fuel governor cut-off point.
For excessive exhaust smoke.
Engine idle speed.
Endorse service record.
Report any additional work required.

**At 30.000 km (18,000 miles) or eighteen months, whichever is first.**

Lubricate all locks (not steering lock), hinges and door — check mechanisms.
Remove road wheels.
Remove road wheel brake drums, wash out dust, inspect shoes for wear and drums for condition.
Inspect wheel cylinders for fluid leaks.
Inspect brake pads for wear, calipers for leaks, and discs for condition.
Refit road wheel brake drums.
Adjust road wheel brakes.
Adjust handbrake if required.
Refit road wheels to original position.
Renew engine oil.
Renew engine oil filter.
Check/top-up gearbox oil.
Check/top-up transfer box oil.
Check/top-up front axle oil.
Check/top-up swivel pin housing oil.
Check/top-up rear axle oil.
Lubricate rear suspension upper link ball joint.
Lubricate propeller shaft universal joints.
Lubricate handbrake mechanical linkage.
Check for oil/fluid leaks from steering and suspension systems.
Check/adjust valve clearance (Turbo diesel only).
Clean/adjust spark plugs.
Check crankcase breathing system for leaks, hoses for security and condition.
Top-up carburetter piston dampers.
Check/top-up fluid in power steering reservoir.

Check/top-up clutch fluid reservoir.
Check/top-up windscreen and rear washer reservoir.
Check power steering system for leaks, hydraulic pipes and unions for chafing and corrosion.
Check condition of driving belts — adjust if required (not diesel camshaft drive belt).
Lubricate distributor (not V8).
Check dwell angle — adjust as necessary (not V8).
Check/adjust ignition timing.

**NOTE:** It is important that the ignition timing, dwell angle and carburetter adjustments are set in accordance with the vehicle engine specification and fuel octane rating. Refer to the relevant repair operation manual for details.

Check/adjust engine idle speed and carburetter mixture settings with engine at normal running temperature.
Check operation of air intake temperature control system (V8).
Check/tighten inlet manifold and exhaust manifold bolts (2.5 litre Turbo-Diesel only).

**Carry out road or roller test.**

 **WARNING: Two wheel roller tests must be restricted to 5 km/hour (3 miles/hour). DO NOT engage the differential lock or the vehicle will drive off the roller test rig because the Land Rover is in permanent four wheel drive.**

Check:
For excessive engine noise.
Clutch for slipping/judder/spinning.
Gear selection/noise — high and low range.
Steering for noise/abnormal effort required.
All instruments, pressure, fuel and temperature gauges, warning indicators.
Heated rear screen.
Shock absorbers (irregularities in ride).
Footbrake, on emergency stop, pulling to one side, binding, pedal effort.
Handbrake efficiency.
Road wheel balance.
Transmission for vibrations.
For body noises (squeaks and rattles).
Fuel governor cut-off point.
For excessive exhaust smoke.
Engine idle speed.
Endorse service record.
Report any additional work required.

**Recommended:** Complete renewal of brake fluid.

# 5 WORKSHOP MAINTENANCE SCHEDULE

**At 40.000 km (24,000 miles) or twenty-four months, whichever is first.**

Check condition and security of seats, seat belt mountings, seat belts and buckles.
Check operation of all lamps.
Check operation of horns.
Check operation of warning indicators.
Check operation of windscreen and rear wipers and washers.
Check condition of wiper blades.
Check security and operation of handbrake.
Check rear view mirror(s) for security, cracks and crazing.
Check operation of all doors, bonnet and tailgate locks.
Check operation of window controls.
Lubricate all locks (not steering lock), hinges and door — check mechanisms.
Lubricate accelerator control linkage and pedal pivot.
Check/adjust tyre pressures including spare.
Check/adjust headlamp alignment.
Check front wheel alignment.
Remove battery connections, clean and grease (refit).
Remove road wheels.
Check tyres comply with Manufacturer's specification.
Check tyres visually for cuts, lumps, bulges, uneven wear and tread depth.
Remove road wheel brake drums, wash out dust, inspect shoes for wear and drums for condition.
Inspect wheel cylinders for fluid leaks.

Inspect brake pads for wear, calipers for leaks, and discs for condition.
Refit road wheel brake drums.
Adjust road wheel brakes.
Adjust handbrake if required.
Refit road wheels to original position.
Renew engine oil.
Renew engine oil filter.
Renew gearbox oil.
Renew transfer box oil.
Renew front axle oil.
Renew swivel pin housing oil.
Renew rear axle oil.
Lubricate rear suspension upper link ball joint.
Lubricate propeller shaft sealed sliding joints
Lubricate propeller shaft universal joints.
Lubricate handbrake mechanical linkage.
Check visually brake, fuel, clutch pipes/unions for chafing, leaks and corrosion.
Check exhaust system for leakage and security.
Check for oil leaks from engine and transmission.
Check for oil/fluid leaks from steering and suspension systems.
Check axle breather pipes, ensure they are not blocked, pinched or split.
Check security and condition of suspension fixings.
Check condition and security of steering unit, joints and gaiters.
Check tightness of propeller shaft coupling bolts.

Clean fuel sedimenter (diesel only).
Renew fuel filter element (petrol).
Drain flywheel housing if drain plug is fitted for wading (refit).
Clean camshaft drive belt housing filter (diesel).
Check condition of heater plug wiring for fraying, chafing and deterioration (diesel only).
Check diesel injectors for correct spray pattern. Ensure no leakage is evident (diesel only).
Renew fuel filter element (diesel).
Check/adjust valve clearance (all models except V8 and Turbo-charged Diesel).
Renew spark plugs.
Renew air cleaner elements.
Check air cleaner dump valve, clean or renew.
Renew engine breather filter (V8).
Clean engine breather filter (all models except V8).
Renew engine flame trap(s) (V8).
Check brake servo hose for security and condition.
Check air injection/pulsair system hoses/pipes for security and condition.
Check operation of pulsair check valves.
Check crankcase breathing system for leaks, hoses for security and condition.
Top-up carburetter piston dampers.
Check/top-up cooling system.

*(continued)*

84

**40.000 km (24,000 miles) or twenty-four months (continued)**

Check/top-up fluid in power steering reservoir.
Check/top-up steering box (manual steering).
Check/adjust steering box.
Check/top-up clutch fluid reservoir.
Check/top-up brake fluid reservoir.
Check/top-up windscreen and rear washer reservoir.
Check cooling and heater system for leaks, hoses for security and condition.
Check power steering system for leaks, hydraulic pipes and unions for chafing and corrosion.
Check condition of driving belts — adjust if required (not camshaft drive belt-diesel).
Check ignition wiring and HT leads for fraying, chafing and deterioration.
Clean distributor cap, check for cracks and tracking.
Clean/adjust distributor points (not V8).
Renew distributor points (not V8).
Lubricate distributor (not V8).
Check voltage drop between coil CB and earth.
Check dwell angle — adjust as necessary (not V8).
Check/adjust ignition timing.

**NOTE:** It is important that the ignition timing, dwell angle and carburetter adjustments are set in accordance with the vehicle engine specification and fuel octane rating. Refer to the relevant repair operation manual for details.

Check throttle operation.
Check/adjust engine idle speed and carburetter mixture settings with engine at normal running temperature.
Check operation of air intake temperature control system (V8).
Check/tighten inlet manifold and exhaust manifold bolts (2.5 litre Turbo-Diesel only).

**Carry out road or roller test:**

 **WARNING: Two wheel roller tests must be restricted to 5 km/hour (3 miles/hour). DO NOT engage the differential lock or the vehicle will drive off the roller test rig because the Land Rover is in permanent four wheel drive.**

Check:
For excessive engine noise.
Clutch for slipping/judder/spinning.
Gear selection/noise — high and low range.
Steering for noise/abnormal effort required.
All instruments, pressure, fuel and temperature gauges, warning indicators.
Heater and air conditioning systems.
Heated rear screen.
Shock absorbers (irregularities in ride).
Footbrake, on emergency stop, pulling to one side, binding, pedal effort.
Handbrake efficiency.

Operation of inertia seat belts.
Road wheel balance.
Transmission for vibration.
For body noises (squeaks and rattles).
Fuel governor cut-off point.
For excessive exhaust smoke.
Engine idle speed.
Endorse service record.
Report any additional work required.

**Recommended:** Where applicable, remove the Pulsair injection manifold and connecting pipes. Ensure that internal bores and cylinder head drillings are clean and free from obstructions. Clean as necessary and refit.
Where the vehicle is operated under severe conditions such as in arid desert and tropical zones, the diesel engine camshaft belt must be renewed.

**At 50.000 km (30,000 miles) or thirty months, whichever is first.**

Lubricate all locks (not steering lock), hinges and door — check mechanisms.
Remove road wheels.
Remove road wheel brake drums, wash out dust, inspect shoes for wear and drums for condition.
Inspect wheel cylinders for fluid leaks.
Inspect brake pads for wear, calipers for leaks, and discs for condition.
Refit road wheel brake drums.
Adjust road wheel brakes.
Adjust handbrake if required.
Refit road wheels to original position.
Renew engine oil.
Renew engine oil filter.
Check/top-up gearbox oil.
Check/top-up transfer box oil.
Check/top-up front axle oil.
Check/top-up swivel pin housing oil.
Check/top-up rear axle oil.
Lubricate rear suspension upper link ball joint.
Lubricate propeller shaft universal joints.
Lubricate handbrake mechanical linkage.
Check for oil/fluid leaks from steering and suspension systems.
Clean/adjust spark plugs.
Check crankcase breathing system for leaks, hoses for security and condition.
Check/adjust valve clearance on Turbo-Diesel.
Top-up carburetter piston dampers.
Check/top-up fluid in power steering reservoir.

Check/top-up clutch fluid reservoir.
Check/top-up brake fluid reservoir.
Check/top-up windscreen and rear washer reservoir.
Check power steering system for leaks, hydraulic pipes and unions for chafing and corrosion.
Check condition of driving belts — adjust if required (not diesel camshaft drive belt).
Lubricate distributor (not V8).
Check dwell angle — adjust as necessary (not V8).
Check/adjust ignition timing.

**NOTE:** It is important that the ignition timing, dwell angle and carburetter adjustments are set in accordance with the vehicle engine specification and fuel octane rating. Refer to the relevant repair operation manual for details.

Check/adjust engine idle speed and carburetter mixture settings with engine at normal running temperature.
Check operation of air intake temperature control system (V8).
Check/tighten inlet manifold and exhaust manifold bolts (2.5 litre Turbo-Diesel only)

**Carry out road or roller test.**

 **WARNING: Two wheel roller tests must be restricted to 5 km/hour (3 miles/hour). DO NOT engage the differential lock or the vehicle will drive off the roller test rig because the Land Rover is in permanent four wheel drive.**

Check:
For excessive engine noise.
Clutch for slipping/judder/spinning.
Gear selection/noise — high and low range.
Steering for noise/abnormal effort required.
All instruments, pressure, fuel and temperature gauges, warning indicators.
Heated rear screen.
Shock absorbers (irregularities in ride).
Footbrake, on emergency stop, pulling to one side, binding, pedal effort.
Handbrake efficiency.
Road wheel balance.
Transmission for vibration.
For body noises (squeaks and rattles).
Fuel governor cut-off point.
For excessive exhaust smoke.
Engine idle speed.
Endorse service record.
Report any additional work required.

**Recommended:** Where the vehicle is operated under adverse conditions such as in dusty atmospheres or high ambient temperatures, the diesel engine camshaft belt must be renewed if not done previously.

**At 60.000 km (36,000 miles) or thirty-six months, whichever is first.**

Check condition and security of seats, seat belt mountings, seat belts and buckles.
Check operation of all lamps.
Check operation of horns.
Check operation of warning indicators.
Check operation of windscreen and rear wipers and washers.
Check condition of wiper blades.
Check security and operation of handbrake.
Check rear view mirror(s) for security, cracks and crazing.
Check operation of all doors, bonnet and tailgate locks.
Check operation of window controls.
Lubricate all locks (not steering lock), hinges and door — check mechanisms.
Lubricate accelerator control linkage and pedal pivot.
Check/adjust tyre pressures including spare.
Check/adjust headlamp alignment.
Check front wheel alignment.
Remove battery connections, clean and grease (refit).
Remove road wheels.
Check tyres comply with Manufacturer's specification.
Check tyres visually for cuts, lumps, bulges, uneven wear and tread depth.
Remove road wheel brake drums, wash out dust, inspect shoes for wear and drums for condition.
Inspect wheel cylinders for fluid leaks.

Inspect brake pads for wear, calipers for leaks, and discs for condition.
Refit road wheel brake drums.
Adjust road wheel brakes.
Adjust handbrake if required.
Refit road wheels to original position.
Renew engine oil.
Renew engine oil filter.
Check/top-up gearbox oil.
Check/top-up transfer box oil.
Check/top-up front axle oil.
Check/top-up swivel pin housing oil.
Check/top-up rear axle oil.
Lubricate rear suspension upper link ball joint.
Lubricate propeller shaft universal joints.
Lubricate handbrake mechanical linkage.
Check visually brake, fuel, clutch pipes/unions for chafing, leaks and corrosion.
Check exhaust system for leakage and security.
Check for oil leaks from engine and transmission.
Check for oil/fluid leaks from steering and suspension systems.
Check axle breather pipes, ensure they are not blocked, pinched or split.
Check security and condition of suspension fixings.
Check condition and security of steering unit, joints and gaiters.
Check tightness of propeller shaft coupling bolts.
Clean fuel sedimenter (diesel only).

Renew fuel filter element (petrol).
Drain flywheel housing if drain plug is fitted for wading (refit).
Clean camshaft drive belt housing filter (diesel).
Check condition of heater plug wiring for fraying, chafing, and deterioration (diesel only).
Renew fuel filter element (diesel).
Check/adjust valve clearance (all models except V8 and Turbo-Diesel).
Renew spark plugs.
Renew air cleaner elements.
Check air cleaner dump valve, clean or renew.
Clean engine breather filter (all models except V8).
Renew engine flame trap(s) (V8).
Renew brake servo filter.
Check brake servo hose for security and condition.
Check V8 air injection/pulsair system hoses/pipes for security and condition.
Check operation of pulsair check valves.
Check crankcase breathing system for leaks, hoses for security and condition.
Top-up carburetter piston dampers.
Check/top-up cooling system.
Check/top-up fluid in power steering reservoir.

*(continued)*

**60.000 km (36,000 miles) or thirty-six months (continued)**

Check/top-up steering box (manual steering).
Check/adjust steering box.
Check/top-up clutch fluid reservoir.
Check/top-up windscreen and rear washer reservoir.
Check cooling and heater system for leaks, hoses for security and condition.
Check power steering system for leaks, hydraulic pipes and unions for chafing and corrosion.
Check condition of driving belts — adjust if required (not camshaft drive belt-diesel).
Check ignition wiring and HT leads for fraying, chafing and deterioration.
Clean distributor cap, check for cracks and tracking.
Clean/adjust distributor points (not V8).
Lubricate distributor (not V8).
Check voltage drop between coil CB and earth.
Check dwell angle — adjust as necessary (not V8).
Check/adjust ignition timing.

**NOTE:** It is important that the ignition timing, dwell angle and carburetter adjustments are set in accordance with the vehicle engine specification and fuel octane rating. Refer to the relevant repair operation manual for details.

Check throttle operation.

Check/adjust engine idle speed and carburetter mixture settings with engine at normal running temperature.
Check operation of air intake temperature control system (V8).
Check/tighten inlet manifold and exhaust manifold bolts (2.5 litre Turbo-Diesel only).

**Carry out road or roller test:**

 **WARNING: Two wheel roller tests must be restricted to 5 km/hour (3 miles/hour). DO NOT engage the differential lock or the vehicle will drive off the roller test rig because the Land Rover is in permanent four wheel drive.**

Check:
For excessive engine noise.
Clutch for slipping/judder/spinning.
Gear selection/noise — high and low range.
Steering for noise/abnormal effort required.
All instruments, pressure, fuel and temperature gauges, warning indicators.
Heater and air conditioning systems.
Heated rear screen.
Shock absorbers (irregularities in ride).
Footbrake, on emergency stop, pulling to one side, binding, pedal effort.
Handbrake efficiency.
Operation of inertia seat belts.

Road wheel balance.
Transmission for vibration.
For body noises (squeaks and rattles).
Fuel governor cut-off point.
For excessive exhaust smoke.
Engine idle speed.
Endorse service record.
Report any additional work required.

**Recommended:** All hydraulic brake fluid, seals and flexible hoses should be renewed, all working surfaces of the master cylinder, wheel cylinders and caliper cylinders should be examined and renewed where necessary. Remove all suspension dampers, test for correct operation, refit or renew as necessary.

At 70.000 km (42,000 miles) or forty-two months, whichever is first.

Lubricate all locks (not steering lock), hinges and door — check mechanisms.
Remove road wheels.
Remove road wheel brake drums, wash out dust, inspect shoes for wear and drums for condition.
Inspect wheel cylinders for fluid leaks.
Inspect brake pads for wear, calipers for leaks, and discs for condition.
Refit road wheel brake drums.
Adjust road wheel brakes.
Adjust handbrake if required.
Refit road wheels to original position.
Renew engine oil.
Renew engine oil filter.
Check/top-up gearbox oil.
Check/top-up transfer box oil.
Check/top-up front axle oil.
Check/top-up swivel pin housing oil.
Check/top-up rear axle oil.
Lubricate rear suspension upper link ball joint.
Lubricate propeller shaft universal joints.
Lubricate handbrake mechanical linkage.
Check for oil/fluid leaks from steering and suspension systems.
Clean/adjust spark plugs.
Check crankcase breathing system for leaks, hoses for security and condition.
Check/adjust valve clearance on Turbo-Diesel.
Top-up carburetter piston dampers.
Check/top-up fluid in power steering reservoir.

Check/top-up clutch fluid reservoir.
Check/top-up brake fluid reservoir.
Check/top-up windscreen and rear washer reservoir.
Check power steering system for leaks, hydraulic pipes and unions for chafing and corrosion.
Check condition of driving belts — adjust if required (not camshaft drive belt — diesel).
Lubricate distributor (not V8).
Check dwell angle — adjust as necessary (not V8).
Check/adjust ignition timing.

NOTE: It is important that the ignition timing, dwell angle and carburetter adjustments are set in accordance with the vehicle engine specification and fuel octane rating. Refer to the relevant repair operation manual for details.

Check/adjust engine idle speed and carburetter mixture settings with engine at normal running temperature.
Check operation of air intake temperature control system (V8).
Check/tighten inlet manifold and exhaust manifold bolts (2.5 litre Turbo-Diesel only)

Carry out road or roller test.

 WARNING: Two wheel roller tests must be restricted to 5 km/hour (3 miles/hour). DO NOT engage the differential lock or the vehicle will drive off the roller test rig because the Land Rover is in permanent four wheel drive.

Check:
For excessive engine noise.
Clutch for slipping/judder/spinning.
Gear selection/noise — high and low range.
Steering for noise/abnormal effort required.
All instruments, pressure, fuel and temperature gauges, warning indicators.
Heated rear screen.
Shock absorbers (irregularities in ride).
Footbrake, on emergency stop, pulling to one side, binding, pedal effort.
Handbrake efficiency.
Road wheel balance.
Transmission for vibrations.
For body noises (squeaks and rattles).
Fuel governor cut-off point.
For excessive exhaust smoke.
Engine idle speed.
Endorse service record.
Report any additional work required.

**At 80.000 km (48,000 miles) or forty-eight months, whichever is first.**

Check condition and security of seats, seat belt mountings, seat belts and buckles.
Check operation of all lamps.
Check operation of horns.
Check operation of warning indicators.
Check operation of windscreen and rear wipers and washers.
Check condition of wiper blades.
Check security and operation of handbrake.
Check rear view mirror(s) for security, cracks and crazing.
Check operation of all doors, bonnet and tailgate locks.
Check operation of window controls.
Lubricate all locks (not steering lock), hinges and door — check mechanisms.
Lubricate accelerator control linkage and pedal pivot.
Check/adjust tyre pressures including spare.
Check/adjust headlamp alignment.
Check front wheel alignment.
Remove battery connections, clean and grease (refit).
Remove road wheels.
Check tyres comply with Manufacturer's specification.
Check tyres visually for cuts, lumps, bulges, uneven wear and tread depth.
Remove road wheel brake drums, wash out dust, inspect shoes for wear and drums for condition.
Inspect wheel cylinders for fluid leaks.

Inspect brake pads for wear, calipers for leaks, and discs for condition.
Refit road wheel brake drums.
Adjust road wheel brakes.
Adjust handbrake if required.
Refit road wheels to original position.
Renew engine oil.
Renew engine oil filter.
Renew gearbox oil.
Renew transfer box oil.
Renew front axle oil.
Renew swivel pin housing oil.
Renew rear axle oil.
Lubricate rear suspension upper link ball joint.
Lubricate propeller shaft sealed sliding joints
Lubricate propeller shaft universal joints.
Lubricate handbrake mechanical linkage.
Check visually brake, fuel, clutch pipes/unions for chafing, leaks and corrosion.
Check exhaust system for leakage and security.
Check for oil leaks from engine and transmission.
Check for oil/fluid leaks from steering and suspension systems.
Check axle breather pipes, ensure they are not blocked, pinched or split.
Check security and condition of suspension fixings.
Check condition and security of steering unit, joints and gaiters.
Check tightness of propeller shaft coupling bolts.

Clean fuel sedimenter (diesel only).
Renew fuel filter element (petrol).
Clean electric fuel pump filter.
Drain flywheel housing if drain plug is fitted for wading (refit).
Clean camshaft drive belt housing filter (diesel).
Remove diesel injectors, clean and reset if necessary.
Check condition of heater plug wiring for fraying, chafing, and deterioration (diesel only).
Check diesel injectors for correct spray pattern. Ensure no leakage is evident (diesel only).
Renew fuel filter element (diesel).
Check/adjust valve clearance (all models except V8 and Turbo-Diesel).
Renew spark plugs.
Renew air cleaner elements.
Remove the Pulsair injection manifold, ensure that the internal bores and the cylinder head drillings are clean and free from obstructions. Clean as necessary and refit. (Emission control V8 petrol engines with carburetters).
Check air cleaner dump valve, clean or renew.
Renew engine breather filter (V8).
Clean engine breather filter (all models except V8).
Renew engine flame trap(s) (V8).

*(continued)*

**80.000 km (48,000 miles) or forty-eight months (continued)**

Check brake servo hose for security and condition.
Check V8 air injection/pulsair system hoses/pipes for security and condition.
Check operation of pulsair check valves.
Check crankcase breathing system for leaks, hoses for security and condition.
Top-up carburetter piston dampers.
Check/top-up cooling system.
Check/top-up fluid in power steering reservoir.
Check/top-up steering box (manual steering).
Check/adjust steering box.
Check/top-up clutch fluid reservoir.
Check/top-up brake fluid reservoir.
Check/top-up windscreen and rear washer reservoir.
Check cooling and heater system for leaks, hoses for security and condition.
Check power steering system for leaks, hydraulic pipes and unions for chafing and corrosion.
Check condition of driving belts — adjust if required (not camshaft drive belt-diesel).
Check ignition wiring and HT leads for fraying, chafing and deterioration.
Clean distributor cap, check for cracks and tracking.
Renew distributor points (not V8).
Lubricate distributor (not V8).
Check voltage drop between coil CB and earth.

Check dwell angle — adjust as necessary (not V8).
Check/adjust ignition timing.

**NOTE:** It is important that the ignition timing, dwell angle and carburetter adjustments are set in accordance with the vehicle engine specification and fuel octane rating. Refer to the relevant repair operation manual for details.

Check throttle operation.
Check/adjust engine idle speed and carburetter mixture settings with engine at normal running temperature.
Check operation of air intake temperature control system (V8).
Check maximum turbo-charge boost pressure (Turbo-Diesel).
Check/tighten inlet manifold and exhaust manifold bolts (2.5 litre Turbo-Diesel only).

**Carry out road or roller test:**

 **WARNING: Two wheel roller tests must be restricted to 5 km/hour (3 miles/hour). DO NOT engage the differential lock or the vehicle will drive off the roller test rig because the Land Rover is in permanent four wheel drive.**

Check:
For excessive engine noise.
Clutch for slipping/judder/spinning.

Gear selection/noise — high and low range.
Steering for noise/abnormal effort required.
All instruments, pressure, fuel and temperature gauges, warning indicators.
Heater and air conditioning systems.
Heated rear screen.
Shock absorbers (irregularities in ride).
Footbrake, on emergency stop, pulling to one side, binding, pedal effort.
Handbrake efficiency.
Operation of inertia seat belts.
Road wheel balance.
Transmission for vibration.
For body noises (squeaks and rattles).
Fuel governor cut-off point.
For excessive exhaust smoke.
Engine idle speed.
Endorse service record.
Report any additional work required.

**Recommended:** Where applicable, remove the Pulsair injection manifold, ensure that the internal bores and cylinder head drillings are clean and free from obstructions. Clean as necessary and refit. Where the vehicle is operated under severe conditions such as in arid desert and tropical zones, the diesel engine camshaft belt must be renewed.

**At 90.000 km (54,000 miles) or fifty-four months, whichever is first.**

Lubricate all locks (not steering lock), hinges and door — check mechanisms.
Remove road wheels.
Remove road wheel brake drums, wash out dust, inspect shoes for wear and drums for condition.
Inspect wheel cylinders for fluid leaks.
Inspect brake pads for wear, calipers for leaks, and discs for condition.
Refit road wheel brake drums.
Adjust road wheel brakes.
Adjust handbrake if required.
Refit road wheels to original position.
Renew engine oil.
Renew engine oil filter.
Check/top-up gearbox oil.
Check/top-up transfer box oil.
Check/top-up front axle oil.
Check/top-up swivel pin housing oil.
Check/top-up rear axle oil.
Lubricate rear suspension upper link ball joint.
Lubricate propeller shaft universal joints.
Lubricate handbrake mechanical linkage.
Check for oil/fluid leaks from steering and suspension systems.
Check/adjust valve clearance on Turbo-charged diesel.
Clean/adjust spark plugs.
Check crankcase breathing system for leaks, hoses for security and condition.
Top-up carburetter piston dampers.
Check/top-up fluid in power steering reservoir.

Check/top-up clutch fluid reservoir.
Completely renew hydraulic brake fluid.
Check/top-up brake fluid reservoir.
Check/top-up windscreen and rear washer reservoir.
Check power steering system for leaks, hydraulic pipes and unions for chafing and corrosion.
Check condition of driving belts — adjust if required (not camshaft drive belt — diesel).
Lubricate distributor (not V8).
Check dwell angle — adjust as necessary (not V8).
Check/adjust ignition timing.

**NOTE:** It is important that the ignition timing, dwell angle and carburetter adjustments are set in accordance with the vehicle engine specification and fuel octane rating. Refer to the relevant repair operation manual for details.

Check/adjust engine idle speed and carburetter mixture settings with engine at normal running temperature.
Check operation of air intake temperature control system (V8).
Check/tighten inlet manifold and exhaust manifold bolts (2.5 litre Turbo-Diesel only)

**Carry out road or roller test:**

 **WARNING: Two wheel roller tests must be restricted to 5 km/hour (3 miles/hour). DO NOT engage the differential lock or the vehicle will drive off the roller test rig because the Land Rover is in permanent four wheel drive.**

Check:
For excessive engine noise.
Clutch for slipping/judder/spinning.
Gear selection/noise — high and low range.
Steering for noise/abnormal effort required.
All instruments, pressure, fuel and temperature gauges, warning indicators.
Heated rear screen.
Shock absorbers (irregularities in ride).
Footbrake, on emergency stop, pulling to one side, binding, pedal effort.
Handbrake efficiency.
Road wheel balance.
Transmission for vibration.
For body noises (squeaks and rattles).
Fuel governor cut-off point.
For excessive exhaust smoke.
Engine idle speed.
Endorse service record.
Report any additional work required.

**At 100.000 km (60,000 miles) or sixty months, whichever is first.**

Check condition and security of seats, seat belt mountings, seat belts and buckles.
Check operation of all lamps.
Check operation of horns.
Check operation of warning indicators.
Check operation of windscreen and rear wipers and washers.
Check condition of wiper blades.
Check security and operation of handbrake.
Check rear view mirror(s) for security, cracks and crazing.
Check operation of all doors, bonnet and tailgate locks.
Check operation of window controls.
Lubricate all locks (not steering lock), hinges and door — check mechanisms.
Lubricate accelerator control linkage and pedal pivot.
Check/adjust tyre pressures including spare.
Check/adjust headlamp alignment.
Check front wheel alignment.
Remove battery connections, clean and grease (refit).
Remove road wheels.
Check tyres comply with Manufacturer's specification.
Check tyres visually for cuts, lumps, bulges, uneven wear and tread depth.
Remove road wheel brake drums, wash out dust, inspect shoes for wear and drums for condition.
Inspect wheel cylinders for fluid leaks.

Inspect brake pads for wear, calipers for leaks, and discs for condition.
Refit road wheel brake drums.
Adjust road wheel brakes.
Adjust handbrake if required.
Refit road wheels to original position.
Renew engine oil.
Renew engine oil filter.
Check/top-up gearbox oil.
Check/top-up transfer box oil.
Check/top-up front axle oil.
Check/top-up swivel pin housing oil.
Check/top-up rear axle oil.
Lubricate rear suspension upper link ball joint.
Lubricate propeller shaft universal joints.
Lubricate handbrake mechanical linkage.
Check visually brake, fuel, clutch pipes/unions for chafing, leaks and corrosion.
Check exhaust system for leakage and security.
Check for oil leaks from engine and transmission.
Check for oil/fluid leaks from steering and suspension systems.
Check axle breather pipes, ensure they are not blocked, pinched or split.
Check security and condition of suspension fixings.
Check condition and security of steering unit, joints and gaiters.
Check tightness of propeller shaft coupling bolts.
Clean fuel sedimenter (diesel only).

Renew fuel filter element (petrol).
Drain flywheel housing if drain plug is fitted for wading (refit).
Clean camshaft drive belt housing filter (diesel).
Check condition of heater plug wiring for fraying, chafing, and deterioration (diesel only).
Renew fuel filter element (diesel).
Check/adjust valve clearance (all models except V8 and Turbo-Diesel).
Renew spark plugs.
Renew air cleaner elements.
Check air cleaner dump valve, clean or renew.
Clean engine breather filter (all models except V8).
Renew engine flame trap(s) (V8).
Check brake servo hose for security and condition.
Check V8 air injection/pulsair system hoses/pipes for security and condition.
Check operation of pulsair check valves.
Check crankcase breathing system for leaks, hoses for security and condition.
Top-up carburetter piston dampers.
Check/top-up cooling system.

*(continued)*

**100.000 km (60,000 miles) or sixty months (continued)**

Check/top-up fluid in power steering reservoir.
Check/top-up steering box (manual steering).
Check/adjust steering box.
Check/top-up clutch fluid reservoir.
Check/top-up brake fluid reservoir.
Check/top-up windscreen and rear washer reservoir.
Check cooling and heater system for leaks, hoses for security and condition.
Check power steering system for leaks, hydraulic pipes and unions for chafing and corrosion.
Check condition of driving belts — adjust if required.
Check ignition wiring and HT leads for fraying, chafing and deterioration.
Clean distributor cap, check for cracks and tracking.
Clean/adjust distributor points (not V8).
Lubricate distributor (not V8).
Check voltage drop between coil CB and earth.
Check dwell angle — adjust as necessary (not V8).
Check/adjust ignition timing.

**NOTE:** It is important that the ignition timing, dwell angle and carburetter adjustments are set in accordance with the vehicle engine specification and fuel octane rating. Refer to the relevant repair operation manual for details.

Check throttle operation.
Check/adjust engine idle speed and carburetter mixture settings with engine at normal running temperature.
Check operation of air intake temperature control system (V8).
Check/tighten inlet manifold and exhaust manifold bolts (2.5 litre Turbo-Diesel only).

**Carry out road or roller test:-**

 **WARNING: Two wheel roller tests must be restricted to 5 km/hour (3 miles/hour). DO NOT engage the differential lock or the vehicle will drive off the roller test rig because the Land Rover is in permanent four wheel drive.**

Check:
For excessive engine noise.
Clutch for slipping/judder/spinning.
Gear selection/noise — high and low range.
Steering for noise/abnormal effort required.
All instruments, pressure, fuel and temperature gauges, warning indicators.
Heater and air conditioning systems.
Heated rear screen.
Shock absorbers (irregularities in ride).
Footbrake, on emergency stop, pulling to one side, binding, pedal effort.
Handbrake efficiency.

Operation of inertia seat belts.
Road wheel balance.
Transmission for vibration.
For body noises (squeaks and rattles).
Fuel governor cut-off point.
For excessive exhaust smoke.
Engine idle speed.
Endorse service record.
Report any additional work required.

**IMPORTANT: Where the vehicle is operated in dusty atmospheres or high ambient temperatures or where the diesel camshaft drive belt has not previously been renewed, it is imperative that this action is taken at this service. Failure to do so could result in serious engine damage.**

**Air cleaner on 4-cylinder petrol and diesel models –
Figs. ST252 and ST253**

**NOTE:** The turbo-diesel air cleaner is illustrated, other models
are similar.

**Renewing the air cleaner element**
Prop open the bonnet, disconnect the air cleaner hoses (2),
pull up the three clips (3) and lift out the air cleaner canister
(4).
Unscrew the wing nut and sealing washer (5) and remove the
integral element and vane assembly (6).
The old element should be discarded and a new one fitted
during reassembly. If a new element is not available, the old
element (if not contaminated with oil or carbon deposits) may
be cleaned by one of the following methods:

**Cleaning the element**
The element can be cleaned either by (A) – the use of
compressed air (which is the better method) or by (B) –
washing.

 **WARNING: Wear safety goggles and breathing mask
during cleaning by compressed air as this method will
create airborne dust and flying particles which can
cause injury.**

**Method (A).** Direct compressed air through the element in the
opposite direction to the normal air flow, keeping the nozzle
at least 25 mm (1 in) from the pleated paper and moving the
nozzle up and down while rotating the element. Maximum air
pressure must be no more than 5,6 kg/cm$^2$ (80 lb/in$^2$) to avoid
damaging the element.

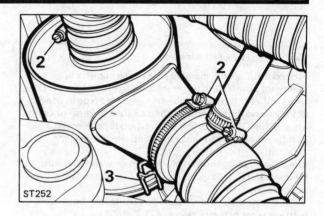

ST252

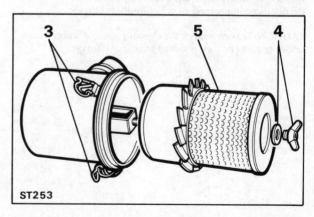

ST253

95

### Cleaning the element (continued)

**Method (B).** Soak the element for 15 to 60 minutes in a detergent powder and water solution. The detergent should be a synthetic, bio-degradable, non-sudsing type as used in automatic washing machines. Brands such as Persil or Omo are suitable. The use of any fluid (especially solvents) or detergents other than those specified may damage the element.

Rinse the element thoroughly. Maximum water pressure must be restricted to 1,7 kg/cm$^2$ (25 lb/in$^2$) to prevent damage to the element.

Allow the element to dry naturally. Drying may be assisted with a maximum temperature of 71°C (160°F).

Do not use compressed air or lightbulbs for drying.

**CAUTION:** Do not replace the element wet as it will collapse under engine suction. resulting in engine damage.

**CAUTION:** Using either of the foregoing methods, care must be taken to prevent dirt being re-deposited on the clean side of the element.

The cleaning instructions given apply only to elements supplied by Land Rover Ltd.

### Inspection of the cleaned element

Using a bright light inside it, rotate the element slowly and inspect it for any rupture, hole, thin spot or other damage. If damage is discovered, the element must be replaced. After inspection, mark the end cap to show the number of times serviced and dates. The element should be replaced after a maximum of six cleanings or annually, whichever occurs first.

### Check air cleaner dump valve

The dump valve provides an automatic drain for the air
cleaner and is fitted in the base of the air cleaner support
bracket.

Squeeze open the dump valve and check that the interior is
clean. Also check that the rubber is flexible and in a good
condition.

If necessary, remove the dump valve to clean the interior. Fit
a new valve if the original is in a poor condition.

### Reassembling

Fit a new element and reassemble the air cleaner.
Replacement procedure is the reverse of removal.

### Turbo-Diesel model

### Air cleaner element change indicator – Fig. LR2052

Located adjacent to the air filter, this indicator clearly shows,
by means of a red band moving across a clear aperture, when
the filter requires changing. Having changed the filter, reset
the indicator by pressing the rounded end until the red band
is no longer visible.

### Air cleaner under heavy conditions

When the vehicle is used in dusty, deep wading or field
conditions, attention to the air cleaner must be more
frequent.

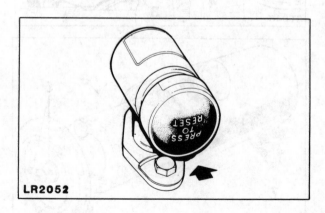

LR2052

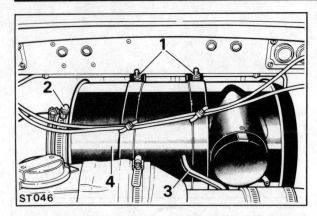

ST046

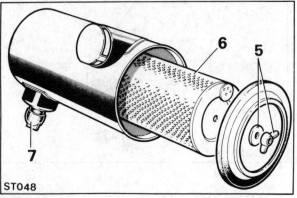

ST048

**Air cleaner (V8 cylinder models) – Figs. ST046 and ST048**
**Removing air cleaner element**
1. Unscrew the two air cleaner strap retaining nuts.
2. Disconnect the air cleaner hose.
3. Remove the engine breather hose.
4. Withdraw air cleaner canister.
5. Unscrew element wing nut and washer and remove filter.
6. Remove the element.

The old element should be discarded and a new one fitted during reassembly. If a new element is not available, it may be possible to clean the old one as described on previous pages.

**Check air cleaner dump valve**
7. Squeeze open the dump valve and check that the interior is clean. Also check that the rubber is flexible and in a good condition.
8. If necessary, remove the dump valve to clean the interior. Fit a new valve if the original is in a poor condition.

**Reassembling**
9. Fit a new element and reassemble the air cleaner.
10. Replacement procedure is the reverse of steps 1 to 4.

**Clean/adjust spark plugs (Petrol models)**

1. The sparking plugs are fitted with plastic covers.
2. To gain access to the plugs for cleaning and gap-setting, pull off the plug covers without detaching them from the high tension leads.
3. Using a spark plug spanner and tommy bar, remove the plugs and washers.
4. Examine the spark plugs. If they are in good condition, clean and adjust as follows:

   Wire-brush the plug threads; open the gap slightly, and vigorously file the electrode sparking surfaces using a point file. This operation is important to ensure correct plug operation by squaring the electrode sparking surfaces.
5. Set the electrode gap to the recommended clearance of 0,71 to 0,84 mm (0.028 to 0.033 in).
6. If satisfactory the plugs and washers may be refitted to the engine but do not overtighten.
7. When pushing the leads on to the plugs, ensure that the shrouds are firmly seated on the plugs.

If new spark plugs are required, use only the type specified in Section 6

Fig. ST051 shows:

A Dirty plug
B Filing plug electrodes
C A clean plug correctly set

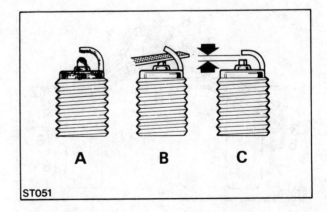

ST051

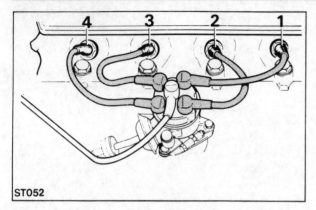

ST052

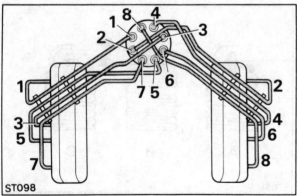

ST098

**Renew spark plugs (Petrol models) – Figs. ST052 and ST098**
To remove spark plugs proceed as follows:
1. Remove the leads from the spark plugs.
2. Using spark plug spanner and tommy bar, remove the plugs and washers.
3. It is important that only spark plugs specified in Data section are used for replacements.
4. Incorrect grades of plug may lead to piston over-heating and engine failure.
5. Wash the new plugs in petrol to remove the protective coating, then set the electrode gaps to the dimension shown in Section 6.
6. Fit the new plugs and washers to the engine but do not overtighten. Push the leads firmly on.

NOTE: The plug leads must be fitted in the order illustrated or the engine will mis-fire. The 4-cylinder engine is illustrated at the top of this page with the V8 below.

### Distributor – (4-cylinder petrol models) – Fig. ST1729

Check and adjust the contact points clearance as follows:

1. Remove the distributor cap and rotor arm; then turn the engine, using the handle, until the contacts are fully open.
2. The clearance should be 0,35 to 0,40 mm (0.014 to 0.016 in) with the feeler gauge a sliding fit between the contacts.
3. If necessary, slacken the screw which secures the adjustable contact.
4. Adjust by the adjuster slot until the clearance is correct; re-tighten the retaining screw.
5. Replace the rotor arm and distributor cap.

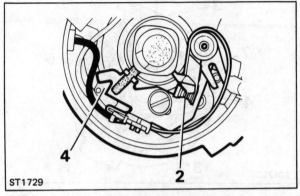

ST1729

### DISTRIBUTOR (4-CYLINDER PETROL MODELS) RENEWING THE CONTACT BREAKER POINTS

#### REMOVE THE OLD CONTACTS

Remove the distributor cap. Remove the rotor arm. Remove the retaining screw and lift the contact set complete from the plate. Press the contact set spring and release the terminal plate and leads from the spring.

#### FIT NEW CONTACTS - ST1082M

Clean the points with petrol to remove the protective coating. Press the contact spring and fit the terminal plate (6) with the black lead uppermost. Fit the contact set to the moving plate, ensuring that the peg (7), underneath the contact pivot, locates in the hole in the moving plate. The sliding contact actuating fork must also locate over the fixed peg. Loosely secure the assembly with the screw, plain and spring washer. Check that the contact leaf spring (10) locates properly in the insulation shoe. Adjust the contact points, as previously described.

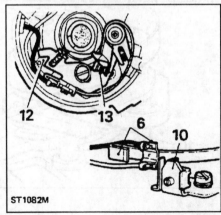

ST1082M

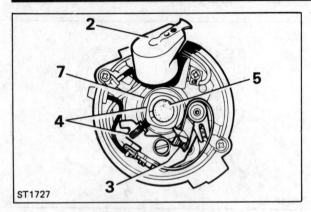

ST1727

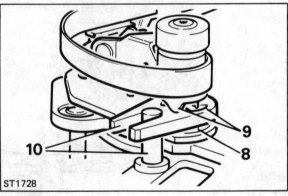

ST1728

**Distributor – clean and lubricate (4-cylinder petrol models)**

Cleaning the points.

1. Release the clips and remove the distributor cap.
2. Pull the rotor arm from the cam spindle.
3. Clean the contact points with fine emery cloth or carborundum stone and wipe clean. Renew the points if worn or pitted.

Lubrication

4. Lightly smear the cam with grease. Do not oil the cam wiping pad.
5. Add a few drops of oil to the felt pad in the top of the cam spindle.
6. Apply a few drops of oil through the gap in the base plate to lubricate the advance mechanism.
7. Every 40,000 km (24,000 miles) add a drop of oil to the moving plate bearing groove.
8. Using grease lubricate the underside of the heel actuator.
9. Grease the actuator ramps and contact breaker heel ribs.
10. Apply grease to the fixed pin and the actuator fork.
11. Align the cam slot and rotor peg and press the rotor arm onto the spindle. Clean the inside of the cap and refit, noting that the cap is located on a peg and can only be fitted one way.

**Electronic ignition (V8 cylinder petrol models) – Fig. RR1249**
A Lucas model 35DM8 distributor is employed. This is an improved design which produces signals from rotating parts, instead of the 'lever' type contact points associated with earlier designs. This results in improved reliability and greatly reduced maintenance.

**Maintenance**
80,000 km (48,000 miles).
Remove the distributor cap and rotor arm and wipe inside with a nap-free cloth.
**Do not disturb** the clear plastic insulating cover which protects the magnetic pick-up module.

RR1249

WARNING: The electronic ignition system involves very high voltages. Inexperienced personnel and wearers of medical pacemaker devices should not be allowed near any part of the high-tension circuit.

Checking of any part of the electronic ignition system must be referred to your Land Rover Dealer or Distributor.

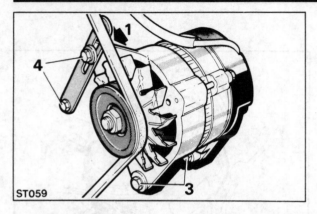

ST059

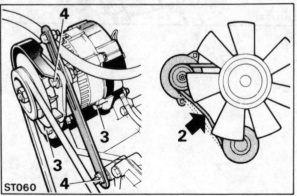

ST060

### Drive belts — general

Examine all pulleys for damage and check there are no
pebbles or grit trapped in the V-grooves that could damage or
reduce the life of the drive belts.

### 4-cylinder models – Fig. ST059

1. Check by thumb pressure between the fan and alternator
   pulleys. Movement should be approximately 9 mm ($\frac{3}{8}$ in).

### V8 cylinder models – Fig. ST060

2. Check by thumb pressure between alternator and
   crankshaft pulleys. Movement should be approximately
   12 mm ($\frac{1}{2}$ in)

### Check fan driving belt, adjust or renew as necessary

Whenever a new fan belt is fitted, re-check deflection after
approximately 1.500 km (1,000 miles) running.

If necessary adjust as follows:

3. Slacken the bolts securing the alternator to the mounting
   bracket.
4. Slacken the fixings at the top and bottom of the
   adjustment link.
5. Pivot the alternator inwards or outwards as necessary and
   adjust until the correct tension is obtained, tighten the
   bolt at the top of the adjustment link.
6. Finally tighten the nut securing the bottom of the
   adjustment link and the two mounting bracket bolts.

**Check driving belt for power steering pump (when fitted) – adjust or renew as necessary**

Whenever a new belt is fitted check adjustment again after approximately 1.500 km (1,000 miles) running.

Check by thumb pressure the belt tension between the crankshaft and pump pulley. Movement should be approximately 12 mm (0.5 in).

If adjustment is necessary:

**4-cylinder models – Fig. ST1730**

1. From underneath the vehicle, slacken the jockey pulley pinch bolt.
2. Pivot the jockey pulley inward or outward, as necessary, to obtain the correct belt tension.
3. Secure the jockey pulley pinch bolt and re-check the tension.

**4-cylinder models - Fig. ST313**

Slacken the pump pivot bolt (1) and the two adjustment clamp bolts (2) and move the pump mounting plate either up or down, as necessary, within the elongated holes, to achieve the correct belt tension.

**V8 cylinder models – Fig. ST091**

1. Slacken the nut on the pivot bolt securing the pump mounting bracket to the cylinder head.
2. Slacken the bolt securing the pump lower bracket to the slotted adjustment link.
3. Slacken the bolt securing the slotted adjustment link to the support bracket mounted on the water pump cover.
4. Pivot the pump as necessary and adjust until the correct belt tension is obtained.
5. Maintaining the tension, tighten the pump adjusting bolts and pivot bolt nut and re-check the tension.

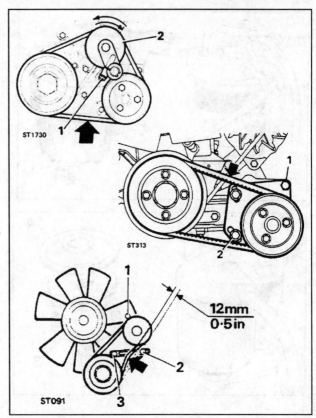

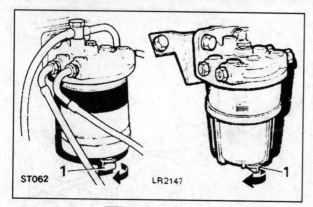

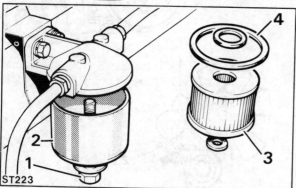

## Fuel filter, paper element type (Diesel models) – Fig. ST062

The filter is mounted at the rear of the engine compartment.
Every day drain off the water as follows:
1. Slacken off drain plug to allow water to run out.
2. When pure diesel fuel is emitted, tighten drain plug.

## Fuel sedimenter, diesel models (when fitted)

The sedimenter is attached to the chassis frame at the side of
the fuel tank. Every day drain off the water as follows:
1. Slacken off drain plug to allow water to run out.
2. When pure diesel fuel is emitted, tighten drain plug.
**NOTE:** If the vehicle is fitted with an extra fuel tank (option), it
may have two sedimenters.

## Renew fuel filter element (Ninety and One Ten Petrol models) – Fig. ST223

The element provides a filter between the pump and
carburetter and is located next to fuel pump on the chassis.
Replace as follows:
1. Unscrew the centre bolt.
2. Withdraw the filter bowl.
3. Remove the small sealing ring and remove element.
4. Withdraw the large sealing ring from the underside of the
   filter body.
5. Discard the old element and thoroughly clean the filter
   bowl.
6. Ensure that the centre and top sealing rings are in good
   condition and replace as necessary.
7. Fit new element, small hole downwards.
8. Refit sealing rings.
9. Replace filter bowl and tighten the centre bolt.

**CLEAN FUEL SEDIMENTER (where fitted) - DIESEL ONLY. Fig. LR2178**
The sedimenter is fitted on the chassis side member, near the rear wheel.

**CLEAN ELEMENT**
Disconnect fuel inlet pipe at sedimenter and raise pipe above level of fuel tank to prevent draining from tank. Support in this position. Support sedimenter bowl (1) and unscrew bolt on top of unit and remove bowl. Remove the sedimenter element (2). Clean all parts in kerosene. Fit new seals (3) and reverse removal procedure. Slacken off the drain plug (4), when pure diesel fuel runs out tighten plug. If necessary, prime the system. Start engine and check for leaks from sedimenter.

**Drain flywheel housing if drain plug is fitted for wading**
**4-cylinder models – Fig. ST064**
**V8 cylinder models – Fig. ST065**

 1. The flywheel housing can be completely sealed to exclude mud and water under severe wading conditions, by fitting a plug in the drain hole at the bottom of the housing.
 2. The plug should only to fitted when the vehicle is expected to do wading or very muddy work.
 3. When the plug is in use it must be removed periodically and all oil allowed to drain off before the plug is replaced.
 4. When the plug is not in use it should be stowed as follows:

**4-cylinder models**

 5. Plug stowed in vehicle tool kit.

**V8 cylinder models**

 6. Plug is screwed into the housing near the drain hole.

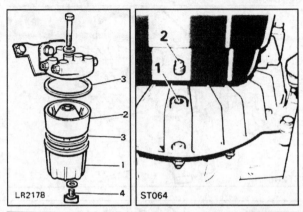

LR2178   ST064

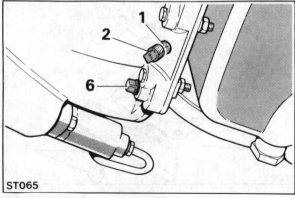

ST065

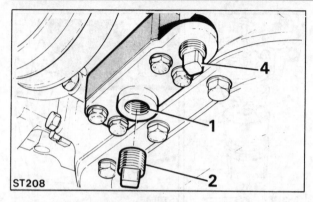

ST208

**Drain engine front timing cover if plug is fitted for wading (Diesel models) – Fig. ST208**

1. The timing cover can be completely sealed to exclude mud and water under severe wading conditions, by fitting a plug in the drain hole at the bottom of the cover.
2. The plug should only be fitted when the vehicle is expected to do wading or very muddy work.
3. When the plug is in use it must be removed periodically to allow any oil to drain off before the plug is replaced.

**NOTE:** There should not be any oil in the timing cover, but if there is, the cause should be investigated as soon as possible, as the timing belt will deteriorate if it becomes contaminated with oil.

4. When the plug is not in use it should be stowed in the tapped hold adjacent to the drain hole.

**Clean filter – engine timing cover (Diesel models) – Fig. ST209**
A gauze filter is fitted at the bottom of the engine timing cover to help prevent mud and other debris entering the drain hole, when the wading plug is not in use. The filter must be removed and cleaned periodically, to ensure that it does not become blocked and prevent the timing cover draining properly. Under normal circumstances, the filter should be cleaned at the intervals specified in the Maintenance Schedule or, more frequently if the vehicle operates regularly in wet or dusty conditions.

1. From underneath the vehicle, remove the four bolts and plain washers and, withdraw the wading plug plate from the bottom of the timing cover.
2. Wash the filter in kerosene or clean fuel. Brush off any mud or other debris and ensure that the whole filter is clean.
3. Check the condition of the gasket for the wading plug plate. If necessary fit a new gasket.
4. Refit the wading plug plate. Tighten the securing bolts.

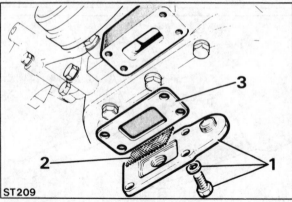

ST209

This symbol may be found on your vehicle or equipment and it means 'CAUTION — do not touch or attempt adjustments until you have read the special instructions concerned on the relevant pages of the Driver's Handbook.'

**WARNING: Some components on your vehicle, such as gaskets and friction surfaces (brake linings or clutch discs), may contain asbestos. Inhaling asbestos dust is dangerous to your health. You are therefore advised to have any maintenance or repair operations on such components carried out by a recognised Land Rover/Range Rover dealer or distributor. If, however, service operations are to be undertaken on parts containing asbestos, the following essential precautions must be observed:**

- **Work out of doors or in a well ventilated area and wear a protective mask.**
- **Dust found on the vehicle or produced during work on the vehicle should be removed by extraction and not by blowing.**
- **Dust waste should be dampened, placed in a sealed container and marked to ensure safe disposal.**
- **If any cutting, drilling etc, is attempted on materials containing asbestos the item should be dampened and only hand tools or low speed power tools used.**

For your further guidance, Land Rover/Range Rover replacement parts which contain asbestos are progressively being identified by the symbol on the left. If you are in any doubt, please consult your dealer or distributor.

The following instructions should be read in conjunction with the brake maintenance recommendations in this Handbook.

**Brake pad replacement**
Your brake pads will require replacement when there is less than 3 mm (0.125 in) of brake lining material remaining. The brake pads fitted to variants with an auxiliary warning system have a built-in electrical sensor to activate the instrument cluster warning light when the pads are worn. If your vehicle has this feature, when purchasing replacement disc pad kits, it is important to ensure that they have sensors and that they have the same friction characteristics.

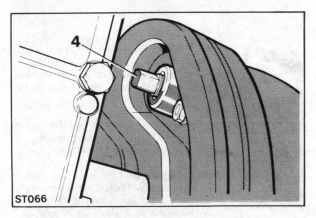

ST066

## Check/adjust transmission handbrake – Fig. ST066

If handbrake movement is excessive, adjust as follows:
1. Set the vehicle on level ground and chock the wheels.
2. Release the handbrake fully.
3. Remove the clevis pin connecting the handbrake lever to the relay at the gearbox end.
4. Fully adjust the handbrake shoe assembly (so that it is fully on) by means of the adjuster on the backplate.
5. Adjust the outer sheath of the handbrake cable by means of the two locknuts at the gearbox end until the holes in the clevis on the inner cable line up with the hole of the relay lever.
6. Fit the clevis pin, washer and a NEW split pin.
7. Slacken the adjuster 1 or 2 notches until handbrake shoes just clear the drum.

8. Apply the handbrake gradually. The drum should still rotate on the first ratchet and start to come on at the second ratchet.

**CAUTION:** DO NOT over adjust the handbrake, the drum must be free to rotate when the handbrake is released, otherwise serious damage will result.

### Foot and handbrake
1. Check operation of foot and handbrake, ensure that the brake pedal travel is not excessive and maintains a satisfactory pressure under normal working load.
2. Excessive pedal travel could be caused by badly worn rear brake linings.
3. If the brakes feel spongy this may be caused by air in the hydraulic system and must be removed by bleeding the system at each wheel cylinder.
4. Prior to this operation, all hydraulic hoses, pipes and connections should be checked for leaks and any leaks rectified.
5. Check operation of handbrake and ensure that it holds the vehicle satisfactorily.

**Check/adjust road wheel brakes**
**Front brake pads – Figs. ST218 and ST067**
The upper illustration shows the front brake for the Land
Rover Ninety; and the lower illustration for the Land Rover
One Ten.

Hydraulic disc brakes are fitted at the front and the correct
brake adjustment is automatically maintained; no provision is
therefore made for adjustment.

1. Check the thickness of the front brake pads and renew if
   the minimum is less than 3,0 mm (0.125 in).
2. Check that rear of brake pad is even across the friction
   face.
3. Check for oil contamination on brake pads and discs, also
   check condition of brake discs for wear and/or corrosion.
4. If replacement or rectification is necessary, this should be
   carried out by your Land Rover Distributor or Dealer.

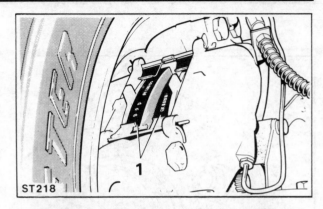

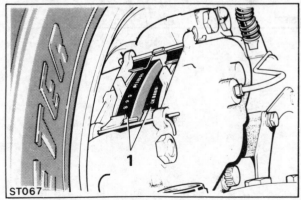

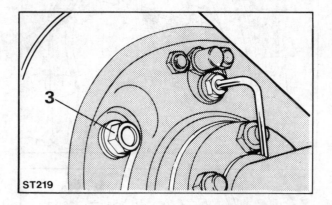

ST219

**Rear brake linings (Ninety models only) – Fig. ST219**

Hydraulic drum brakes are fitted at the rear and require the following attention.

When the vehicle is used in deep muddy conditions the brake drums must be periodically removed and cleaned, at the same time the brake shoes and anchor plate should be thoroughly cleaned.

When used continuously under exceptionally wet and muddy conditions this operation may be advisable once, or even twice a week, to prevent the abrasive action of packed mud rapidly wearing out brake linings and drums.

When lining wear has reached the point where the pedal travel becomes excessive, it is necessary to adjust the brake shoes closer to the drum.

Proceed as follows:

1. The shoes are set by a single hexagon adjustment bolt operating through a serrated snail cam enabling both shoes to be adjusted to obtain the best results.
2. Jack up one rear wheel.
3. Check that the raised wheel rotates freely then turn the adjuster until the brake shoe is in firm contact with the drum.
4. Slacken off the adjuster just sufficiently for the drum to rotate freely.
5. Lower the wheel to the ground.
6. Repeat the procedure for the other wheel.

**Rear brake linings (One Ten models only) – Fig. ST068**
Hydraulic drum brakes are fitted at the rear and require the following attention.

When the vehicle is used in deep muddy conditions the brake drums must be periodically removed and cleaned, at the same time the brake shoes and anchor plate should be thoroughly cleaned.

When used continuously under exceptionally wet and muddy conditions this operation may be advisable once, or even twice a week, to prevent the abrasive action of packed mud rapidly wearing out brake linings and drums.

When lining wear has reached the point where the pedal travel becomes excessive, it is necessary to adjust the brake shoes closer to the drum.

Proceed as follows:

1. Each shoe is independently set by means of a hexagon adjustment bolt operating through a serrated snail cam and each shoe should be set individually to obtain the best results.
2. Jack up one rear wheel.
3. Check that the raised wheel rotates freely then turn one adjuster until the brake shoe is in firm contact with the drum.
4. Slacken off the adjuster just sufficiently for the drum to rotate freely.
5. Repeat for the other brake shoe.
6. Lower the wheel to the ground.
7. Repeat the procedure for the other wheel.

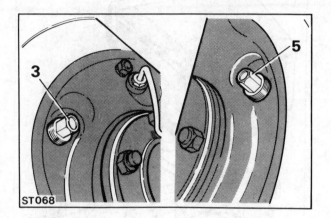

ST068

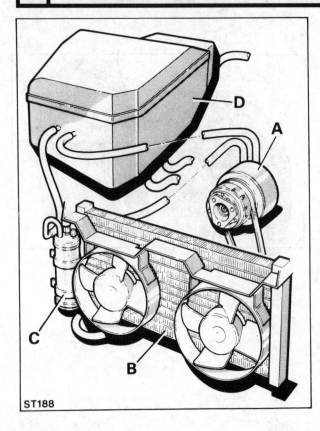

ST188

**Air conditioning system (option) – Fig. ST188**

The air conditioning system operates in conjunction with the vehicle heater to provide cooled and dried recirculated or fresh air.

The system is made up of four separate units.

(A)  An engine-mounted compressor.

(B)  A condenser mounted in front of the radiator.

(C)  A receiver/drier unit located in the engine compartment.

(D)  An evaporator-heater unit mounted in the engine compartment.

The four units are interconnected by hoses carrying refrigerant. The refrigerant circuit cools the evaporator which is connected to the ventilation system, and thus cools the air as it enters the vehicle.

The system delivers hot, cooled, fresh, recirculated and dehumidified air as required to all positions.

The installation incorporates temperature, fan speed and distribution controls mounted on the fascia.

⚠️ **WARNING: The air conditioning system is filled at high pressure with a potentially toxic material. Follow service instructions when dismantling or applying excessive heat, e.g. steam cleaning, painting, etc. Servicing must only be carried out by a qualified engineer in accordance with instructions in the Repair Operation Manual.**

**Condenser**
Using a water hose or air line, clean the exterior of the
condenser matrix.
Check the pipe connections for signs of fluid leakage.

**Evaporator**
Examine the pipe connections for signs of fluid leakage.

**Receiver/drier**
Check the pipe connections for signs of fluid leakage.

**Compressor**
Check the pipe connections for fluid leakage and hoses for
swelling.

**Recommended refrigerants and oils**
See Data Section 6.

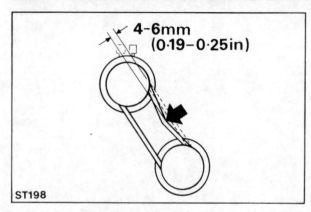

ST198

### Compressor drive-belt – Fig. ST198

The belt must be adjusted with not more than 4 to 6 mm (0.19 to 0.25 in) total deflection when checked by hand mid-way between the pulleys on the longest run.

Where the belt has stretched beyond the limits, a noisy whine or knock will be evident during operation.

If necessary, adjust as follows:

### Belt adjustment (4-cylinder Petrol and Diesel models) – Fig. ST203

Slacken all adjustment bolts associated with the compressor (1) and the lower pulley pivot fixing (2).

Adjust the position of the lower pulley to give a belt tension of 4 to 6 mm (0.19 to 0.25 in). Tighten the pivot bolt and recheck the tension.

Alter the position of the compressor to give a belt tension of 4 to 6 mm (0.19 to 0.25 in). Secure all compressor adjustment bolts and recheck the tension.

### Belt adjustment (V8 Petrol model)

Slacken the compressor adjuster bolts.

Adjust the position of the compressor to give the correct belt tension of 4 to 6 mm (0.19 to 0.25 in).

Tighten all fixings and recheck the belt tension.

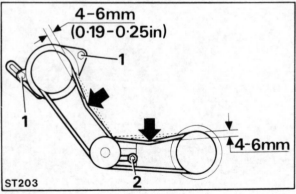

ST203

**Engine, 4-cylinder petrol models**

| | |
|---|---|
| Bore | 90,47 mm (3.562 in) |
| Stroke | 97,0 mm (3.819 in) |
| Number of cylinders | 4 |
| Cylinder capacity | 2495 cc (152.2 cu in) |
| Compression ratio | 8.0:1 |
| Firing order | 1, 3, 4, 2 |
| Sparking plug type | Champion N9YC |
| Sparking plug point gap | 0,72 to 0,88 mm (0.028 to 0.035 in) |
| Distributor contact breaker gap | 0,35 to 0,40 mm (0.014 to 0.016 in) |
| Dwell angle | 49° to 59° |
| Ignition timing, dynamic; models with emission control | 16° BTDC at 2000 rpm with vacuum pipe disconnected when using 90 octane fuel — 2 star rating in UK |

In an emergency where dynamic check equipment is not available, the ignition timing can be set statically at TDC. It should be checked and adjusted dynamically as soon as possible

| | |
|---|---|
| Tappet clearance, inlet | 0,25 mm (0.010 in)  ⎱ Engine at |
| Tappet clearance, exhaust | 0,25 mm (0.010 in)  ⎰ running temperature |
| Valve timing (No. 1 exhaust valve peak) | 104° BTDC |
| Carburetter | Weber 32/34 DMTL |
| Oil pressure | 2.5 to 4,5 kgf/cm$^2$ (35 to 65 lbf/in$^2$) at 50 kph (30 mph) in top gear with engine warm |

**Engine, 4-cylinder diesel models**

| | |
|---|---|
| Bore | 90,47 mm (3.562 in) |
| Stroke | 97,0 mm (3.819 in) |
| Number of cylinders | 4 |
| Compression ratio | 21.0:1 |
| Cylinder capacity | 2495 cc (152 cu in) |
| Firing order | 1, 3, 4, 2 |
| Injection timing | Crankshaft at EP, set injection pump using special tool 18G 1458 |
| Tappet clearance, inlet | 0,25 mm (0.010 in)   ⎫   Engine hot |
| Tappet clearance, exhaust | 0,25 mm (0.010 in)   ⎬   or cold |
| Valve timing (No. 1 exhaust valve peak) | 106° to 109° |
| Oil pressure | 2,5 to 4,5 kgf/cm² (35 to 65 lbf/in²) at 50 kph (30 mph) in top gear with engine warm |

**Main gearbox – 4-cylinder petrol and diesel models**

| | | |
|---|---|---|
| Type — Manual | 5-speed helical constant mesh, with synchromesh on all forward gears | |
| Main gearbox ratios | Fifth (Cruising gear) | 0.83:1 |
| | Fourth | 1.00:1 |
| | Third | 1.51:1 |
| | Second | 2.30:1 |
| | First | 3.58:1 |
| | Reverse | 3.70:1 |

(continued)

## Transfer gearbox – 4-cylinder petrol and diesel models

| | | |
|---|---|---|
| Type.................................................................... | LT230T. Two-speed reduction on main gearbox output. Front and rear drive permanently engaged via a lockable differential. | |
| — Ninety models................................ | High | 1.4109:1 |
| | Low | 3.3198:1 |
| — One Ten models............................ | High | 1.6670:1 |
| | Low | 3.3198:1 |

## Rear axle

| | |
|---|---|
| Type — Ninety models .................................................. | Spiral bevel |
| Type — One Ten models ............................................... | Hypoid; full floating shafts, Salisbury 8HA |
| Ratio — All models ...................................................... | 3.54:1 |

## Front axle

| | |
|---|---|
| Differential................................................................. | Spiral bevel |
| Front wheel drive ....................................................... | Enclosed constant velocity joint |
| Ratio ........................................................................ | 3.54:1 |

Overall ratio (including final drive) — Ninety models

| | In high transfer | In low transfer |
|---|---|---|
| Fifth (Cruising gear) | 4.15:1 | 9.76:1 |
| Fourth | 4.99:1 | 11.75:1 |
| Third | 7.53:1 | 17.71:1 |
| Second | 11.49:1 | 27.03:1 |
| First | 17.90:1 | 42.11:1 |
| Reverse | 18.48:1 | 43.47:1 |

Overall ratio (including final drive) — One Ten models

| | In high transfer | In low transfer |
|---|---|---|
| Fifth (Cruising gear) | 4.90:1 | 9.76:1 |
| Fourth | 5.89:1 | 11.75:1 |
| Third | 8.89:1 | 17.71:1 |
| Second | 13.57:1 | 27.03:1 |
| First | 21.15:1 | 42.11:1 |
| Reverse | 21.83:1 | 43.47:1 |

**Engine — V8 cylinder models**

| | |
|---|---|
| Bore | 88,9 mm (3.500 in) |
| Stroke | 71,12 mm (2.800 in) |
| Number of cylinders | 8 |
| Cylinder capacity | 3528 cc (215 cu in) |
| Compression ratio | 8.13:1 |
| Firing order | 1, 8, 4, 3, 6, 5, 7, 2 |
| Sparking plug type | Champion N12Y |
| Sparking plug gap | 0,71 to 0,84 mm (0.028 to 0.033 in) |
| Distributor | Lucas 35DM8. Electronic |
| Ignition timing, dynamic; models with emission control | 6° BTDC with vacuum pipes connected using 91 - 93 octane fuel — 2 star rating in UK |
| Ignition timing, dynamic; non-emission engines | 6° BTDC at 700 rpm maximum with vacuum pipe connected using 91 - 93 octane fuel |
| Carburetters | Twin S.U. type H.I.F. 44 |
| Oil pressure | 2,1 to 2,8 kgf/cm² (30 to 40 lbf /in²) at 80 kph (50 mph) in top gear with engine warm |

**Main gearbox — V8 models**

| | | |
|---|---|---|
| Type — Manual | LT 85. Five-speed helical constant mesh, with synchromesh on all forward gears | |
| Main gearbox ratios | Fifth (Cruising gear) | 0.7951:1 |
| | Fourth | 1.0000:1 |
| | Third | 1.4362:1 |
| | Second | 2.1804:1 |
| | First | 3.6497:1 |
| | Reverse | 3.718:1 |

**Transfer gearbox — V8 models**

Type........................................................................ LT230T. Two-speed reduction on main gearbox output. Front
and rear drive permanently engaged via a lockable differential.

| | — Ninety models | High | 1.222:1 |
|---|---|---|---|
| | | Low | 3.3198:1 |
| | — One Ten models | High | 1.410:1 |
| | | Low | 3.3198:1 |

**Rear axle**

Type — Ninety models ................................................. Spiral bevel
Type — One Ten models .............................................. Hypoid; full floating shafts, Salisbury 8HA
Ratio — All models ...................................................... 3.5385:1

**Front axle**

Differential............................................................... Spiral bevel
Front wheel drive ...................................................... Enclosed constant velocity joint
Ratio ......................................................................... 3.5385:1

Overall ratio (including final drive) — Ninety models

| | In high transfer | In low transfer |
|---|---|---|
| Fifth (Cruising gear) | 3.4380:1 | 9.3401:1 |
| Fourth | 4.3240:1 | 11.7471:1 |
| Third | 6.2102:1 | 16.8712:1 |
| Second | 9.4282:1 | 25.6134:1 |
| First | 15.7815:1 | 42.8734:1 |
| Reverse | 16.0768:1 | 43.6758:1 |

Overall ratio (including final drive) — One Ten models

| | In high transfer | In low transfer |
|---|---|---|
| Fifth (Cruising gear) | 3.9695:1 | 9.3401:1 |
| Fourth | 4.9925:1 | 11.7471:1 |
| Third | 7.1702:1 | 16.8712:1 |
| Second | 10.8856:1 | 25.6134:1 |
| First | 18.2210:1 | 42.8734:1 |
| Reverse | 18.5620:1 | 43.6758:1 |

**Steering (lock to lock)**

Manual ...................................................................... 4.4 turns
Power assisted ........................................................... 3.0 turns

## Vehicle Dimensions — Ninety models

| | | Soft Top | | | Pick-up | | | Hard Top | | | Station Wagon | | |
|---|---|---|---|---|---|---|---|---|---|---|---|---|---|
| | | 2.5P | 3.5P | 2.5D | 2.5P | 3.5P | 2.5D | 2.5P | 3.5P | 2.5D | 2.5P | 3.5P | 2.5D |
| **DIMENSIONS** | | | | | | | | | | | | | |
| Overall Length | mm (in) | 3722 (146.5) | | | 3722 (146.5) | | | 3883 (152.9) | | | 3883 (152.9) | | |
| Overall Width | mm (in) | 1790 (70.5) | | | 1790 (70.5) | | | 1790 (70.5) | | | 1790 (70.5) | | |
| 2400kg Height† | mm (in) | 1965 (77.4) | | | 1963 (77.3) | | | 1972 (77.6) | | | 1963 (77.3) | | |
| 2550kg Height† | mm (in) | 2000 (78.7) | | | 1993 (78.5) | | | 1997 (78.6) | | | 1989 (78.3) | | |
| Wheelbase | mm (in) | 2360 (92.9) | | | 2360 (92.9) | | | 2360 (92.9) | | | 2360 (92.9) | | |
| Track Front/Rear | mm (in) | 1486 (58.5) | | | 1486 (58.5) | | | 1486 (58.5) | | | 1486 (58.5) | | |
| Cargo Bed Length | mm (in) | 1144 (45.0) | | | 1144 (45.0) | | | 1144 (45.0) | | | 1144 (45.0) | | |
| Interior Width | mm (in) | 1620 (63.8) | | | 1620 (63.8) | | | 1620 (63.8) | | | 1620 (63.8) | | |
| Interior Height | mm (in) | 1215 (47.8) | | | — | | | 1215 (47.8) | | | 1215 (47.8) | | |
| Width between Wheel Boxes | mm (in) | 925 (36.4) | | | 925 (36.4) | | | 925 (36.4) | | | 925 (36.4) | | |
| Seating Capacity | | 2—7 | | | 2—7 | | | 2—7 | | | 6—7 | | |
| **PERFORMANCE** | | | | | | | | | | | | | |
| Tyre size fitted | | 6.00 × 16 | | | 7.50 × 16 (except XS) | | | 205 × 16 | | | | | |
| Min. Turning Radius (kerb to kerb) m (ft) | | 5,75 (18.9) | | | 6,15 (20.2) | | | 5,85 (19.2) | | | | | |
| Max. Gradient (EEC kerb weight) | | 45° | | | 45° | | | 45° | | | | | |
| Approach Angle (EEC kerb weight) | | 47° | | | 51° | | | 48° | | | | | |
| Departure Angle (EEC kerb weight) | | 48° | | | 52° | | | 48° | | | | | |
| Ramp Break Over Angle | | 149° | | | 146° | | | 150° | | | | | |
| Min. Ground Clearance (unladen) mm (in) | | 198 (7.8) | | | 229 (9) | | | 191 (7.5) | | | | | |
| Wading Depth | mm (in) | 500 (20) | | | 500 (20) | | | 500 (20) | | | | | |
| **TOWING WEIGHTS** | Towing Weights | 2.5 PETROL | | | 3.5 PETROL | | | | 2.5 DIESEL | | | | |
| | Unbraked Trailers | 750kg | | | 750kg | | | | 750kg | | | | |
| | Trailers with Over Run Brakes | 3500kg | | | 3500kg | | | | 3500kg | | | | |
| | 4-wheel Trailers with Coupled Brakes | 4000kg | | | 4000kg | | | | 3500kg | | | | |

†Height depends upon suspension and tyres specified.

**Vehicle Dimensions — One Ten models**

| | | Soft Top | | | Pick-up | | | Hard Top | | | Station Wagon | | | High Capacity Pick-up | | |
|---|---|---|---|---|---|---|---|---|---|---|---|---|---|---|---|---|
| | | 2.5P | 2.5D | 3.5P | 2.5P | 2.5D | 3.5P | 2.5P | 2.5D | 3.5P | 2.5P | 2.5D | 3.5P | 2.5P | 2.5D | 3.5P |
| **DIMENSIONS** Overall Length | mm (in) | | 4438 (175) | | | 4438 (175) | | | 4599 (181.1) | | | 4599 (181.1) | | | 4631 (182) | |
| Overall Width | mm (in) | | 1790 (70.5) | | | 1790 (70.5) | | | 1790 (70.5) | | | 1790 (70.5) | | | 1790 (70.5) | |
| Overall Height† | mm (in) | | 2035 (80.1) | | | 2035 (80.1) | | | 2035 (80.1) | | | 2035 (80.1) | | | 2035 (80.1) | |
| Wheelbase | mm (in) | | 2794 (110) | | | 2794 (110) | | | 2794 (110) | | | 2794 (110) | | | 2794 (110) | |
| Track Front/Rear | mm (in) | | 1486 (58.5) | | | 1486 (58.5) | | | 1486 (58.5) | | | 1486 (58.5) | | | 1486 (58.5) | |
| Cargo Bed Length | mm (in) | | 1900 (74.8) | | | 1900 (74.8) | | | 1900 (74.8) | | | — | | | 2010 (79.2) | |
| Interior Width | mm (in) | | 1620 (63.8) | | | 1620 (63.8) | | | 1620 (63.8) | | | 1620 (63.8) | | | 1660 (65.3) | |
| Interior Height | mm (in) | | 1205 (47.4) | | | — | | | 1205 (47.4) | | | 1205 (47.4) | | | — | |
| Width between Wheelboxes | mm (in) | | 925 (36.4) | | | 925 (36.4) | | | 925 (36.4) | | | 925 (36.4) | | | 1090 (43) | |
| Seating Capacity | | | 2—3—11 | | | 2—3—11 | | | 2—3—11 | | | 9—10—11—12 | | | 2—3 | |

| | | |
|---|---|---|
| **PERFORMANCE** Min. Turning Radius | m (ft) | 6,4 (21) |
| Max. Gradient | | 45°max |
| Approach Angle | | 50° (at EEC kerb weight) |
| Departure Angle | | 35° (at EEC kerb weight) |
| Ramp Break Over Angle | | 152° |
| Min. Ground Clearance | mm (in) | 215 (8.5) |
| Wading Depth | mm (in) | 500 (20) |

| **TOWING WEIGHTS** Towing Weights | 2.3 PETROL | 3.5 PETROL | 2.5 DIESEL |
|---|---|---|---|
| Unbraked Trailers | 750kg | 750kg | 750kg |
| Trailers with Over Run Brakes | 3500kg | 3500kg | 3500kg |
| 4-wheel Trailers with coupled brakes  FULLY BRAKED | 4000kg | 4000kg | 3500kg |

†Height depends upon suspension and tyres specified.
**NOTE:** All weight figures quoted are subject to local legal restrictions.

## Vehicle Weights and Payload

Payload figures quoted in the accompanying table are nominal values for a base specification vehicle and will in general represent the maximum, as any options or extras fitted to the vehicle will increase its unladen weight and hence decrease its allowable payload.

When loading a vehicle to its maximum (Gross Vehicle Weight), consideration must be taken of the unladen vehicle weight and the distribution of the payload to ensure that axle loadings do not exceed the permitted maximum values.

It is the customer's responsibility to limit the vehicle's payload in an appropriate manner such that neither maximum axle loads nor Gross Vehicle Weight are exceeded.

| Land Rover Ninety | | Soft Top | | | Pick-up | | | Hard Top | | | Station Wagon | | |
|---|---|---|---|---|---|---|---|---|---|---|---|---|---|
| **Model — Petrol/Diesel** | | 2.5P | 2.5D | 3.5P | 2.5P | 2.5D | 3.5P | 2.5P | 2.5D | 3.5P | 2.5P | 2.5D | 3.5P |
| Gross Vehicle Weight | STANDARD SUSPENSION 2400 kg | | | | | | | | | | | | |
| EEC Kerb Weight | kg | 1606 | 1643 | 1602 | 1635 | 1672 | 1631 | 1648 | 1685 | 1644 | 1690 | 1727 | 1686 |
| EEC Payload | kg | 794 | 757 | 798 | 765 | 728 | 769 | 752 | 715 | 756 | 710 | 673 | 714 |
| Unladen Weight | kg | 1487 | 1519 | 1483 | 1516 | 1548 | 1512 | 1529 | 1561 | 1525 | 1571 | 1603 | 1567 |
| Payload | kg | 913 | 881 | 917 | 884 | 852 | 888 | 871 | 839 | 875 | 829 | 797 | 833 |
| Maximum Axle Weights, all Ninety models with Standard Suspension Front Axle 1200 kg Rear Axle 1380 kg | | | | | | | | | | | | | |
| Gross Vehicle Weight | HIGH LOAD SUSPENSION 2550 kg | | | | | | | | | | | | |
| EEC Kerb Weight | kg | 1633 | 1670 | 1629 | 1662 | 1699 | 1658 | 1675 | 1712 | 1671 | 1717 | 1754 | 1713 |
| EEC Payload | kg | 917 | 880 | 921 | 888 | 851 | 892 | 875 | 838 | 879 | 833 | 796 | 837 |
| Unladen Weight | kg | 1514 | 1546 | 1510 | 1543 | 1575 | 1539 | 1556 | 1588 | 1522 | 1598 | 1630 | 1594 |
| Payload | kg | 1036 | 1004 | 1040 | 1007 | 975 | 1011 | 994 | 962 | 998 | 952 | 920 | 956 |
| Maximum Axle Weights, all Ninety models with High Load Suspension Front Axle 1200 kg Rear Axle 1500 kg | | | | | | | | | | | | | |

**Vehicle Weights and Payload**

Payload figures quoted in the accompanying table are nominal values for a base specification vehicle and will in general represent the maximum, as any options or extras fitted to the vehicle will increase its unladen weight and hence decrease its allowable payload.

When loading a vehicle to its maximum (Gross Vehicle Weight), consideration must be taken of the unladen vehicle weight and the distribution of the payload to ensure that axle loadings do not exceed the permitted maximum values.

It is the customer's responsibility to limit the vehicle's payload in an appropriate manner such that neither maximum axle loads nor Gross Vehicle Weight are exceeded.

| Land Rover One Ten | | Soft Top | | | Pick-up | | | Hard Top | | | Station Wagon | | | High Capacity Pick-up | | |
|---|---|---|---|---|---|---|---|---|---|---|---|---|---|---|---|---|
| Model — Petrol/Diesel | | 2.5P | 2.5D | 3.5P | 2.5P | 2.5D | 3.5P | 2.5P | 2.5D | 3.5P | 2.5P | 2.5D | 3.5P | 2.5P | 2.5D | 3.5P |
| Gross Vehicle Weight | UNLEVELLED SUSPENSION 3050 kg | | | | | | | | | | | | | | | |
| EEC Kerb Weight | kg | 1723 | 1742 | 1698 | 1724 | 1743 | 1699 | 1777 | 1796 | 1752 | 1887 | 1906 | 1862 | 1813 | 1859 | 1778 |
| EEC Payload | kg | 1327 | 1308 | 1352 | 1326 | 1307 | 1351 | 1273 | 1254 | 1298 | 1163 | 1144 | 1188 | 1237 | 1191 | 1272 |
| Unladen Weight | kg | 1588 | 1599 | 1563 | 1589 | 1600 | 1564 | 1642 | 1653 | 1617 | 1752 | 1763 | 1727 | 1678 | 1716 | 1643 |
| Payload | kg | 1462 | 1451 | 1487 | 1461 | 1450 | 1486 | 1408 | 1397 | 1433 | 1298 | 1287 | 1323 | 1372 | 1334 | 1407 |
| Maximum Axle Weights, all One Ten models with Unlevelled Suspension | | | | | | | | | | | | | | | | |
| Front Axle 1200 kg    Rear Axle 1850 kg | | | | | | | | | | | | | | | | |
| Gross Vehicle Weight | LEVELLED SUSPENSION 2950 kg | | | | | | | | | | | | | | | |
| EEC Kerb Weight | kg | 1733 | 1752 | 1708 | 1734 | 1753 | 1709 | 1787 | 1806 | 1762 | 1897 | 1916 | 1872 | 1823 | 1869 | 1788 |
| EEC Payload | kg | 1217 | 1198 | 1242 | 1216 | 1197 | 1241 | 1163 | 1144 | 1188 | 1053 | 1034 | 1078 | 1127 | 1081 | 1162 |
| Unladen Weight | kg | 1598 | 1609 | 1573 | 1599 | 1610 | 1574 | 1652 | 1663 | 1627 | 1762 | 1773 | 1737 | 1688 | 1726 | 1653 |
| Payload | kg | 1352 | 1341 | 1377 | 1351 | 1340 | 1376 | 1298 | ·1287 | 1323 | 1188 | 1177 | 1213 | 1262 | 1224 | 1297 |
| Maximum Axle Weights, all One Ten models with Levelled Suspension | | | | | | | | | | | | | | | | |
| Front Axle 1200 kg    Rear Axle 1750 kg | | | | | | | | | | | | | | | | |

**Land Rover 127" Crew Cab**

| | | 2.5 Petrol | | 2.5 Diesel | | 3.5 Petrol | |
|---|---|---|---|---|---|---|---|
| | | Unlevelled | Levelled | Unlevelled | Levelled | Unlevelled | Levelled |
| Gross Vehicle Weight | Front Axle | 1200 | 1200 | 1200 | 1200 | 1200 | 1200 |
| | Rear Axle | 1850 | 1750 | 1850 | 1750 | 1850 | 1750 |
| | Total | 3050 | 2950 | 3050 | 2950 | 3050 | 2950 |
| E.E.C. Kerb Weight | Front Axle | 1062 | 1067 | 1070 | 1075 | 1027 | 1027 |
| | Rear Axle | 1002 | 1017 | 1015 | 1025 | 985 | 995 |
| | Total | 2064 | 2074 | 2085 | 2095 | 2012 | 2022 |
| E.E.C. Payload | Total | 986 | 876 | 965 | 855 | 1038 | 928 |
| Unladen Weight | Total | 1924 | 1934 | 1936 | 1946 | 1872 | 1882 |
| Payload | Total | 1126 | 1016 | 1114 | 1004 | 1178 | 1068 |
| E.E.C. Kerb Weight Plus Five Passengers | Total | 2439 | 2449 | 2460 | 2470 | 2387 | 2397 |

**NOTE:** All other vehicle details, (including tyre pressures) are the same as those given for the Land Rover 110" in this Driver's Handbook.

## Capacities

The following capacity figures are approximate and are provided as a guide only. All oil level must be set using the dipstick or level plugs as applicable. Refer to Section 4 for the correct procedure for checking the engine sump.

| Component | Litres | Imperial unit |
|---|---|---|
| Engine sump oil (4-cylinder) | 6,00 | 10.56 pints |
| Extra when refilling after fitting new filter (4-cylinder) | 0,85 | 1.50 pints |
| Engine sump oil (V8 cylinder) | 5,10 | 9.00 pints |
| Extra when refilling after fitting new filter (V8 cylinder) | 0,56 | 1.00 pint |
| Main gearbox oil (LT77) 4-cylinder | 2,20 | 3.90 pints |
| Main gearbox oil (LT85) V8 cylinder | 3,00 | 5.28 pints |
| Transfer box oil, all models | 2,80 | 4.90 pints |
| Front differential | 1,70 | 3.00 pints |
| Rear differential (Ninety models) | 1,70 | 3.00 pints |
| Rear differential: Salisbury 8HA (One Ten models) | 2,26 | 4.00 pints |
| Steering box - manual | 0,43 | 0.75 pint |
| Power steering box and reservoir | 2,90 | 5.00 pints |
| Swivel pin housing oil (each) | 0,35 | 0.60 pint |
| Fuel tank, rear (One Ten models) | 79,50 | 17.50 gallons |
| Fuel tank, side (except One Ten Station wagon) | 68,20 | 15.00 gallons |
| Fuel tank, side (One Ten Station wagon only) | 45,50 | 10.00 gallons |
| Fuel tank, side (Ninety models) | 54,58 | 12.00 gallons |
| Cooling system, 4-cylinder petrol models and naturally aspirated diesel models | 10,8 | 19.0 pints |
| Cooling system, 4-cylinder diesel models and heavy duty petrol models | 10,80 | 19.00 pints |
| Cooling system, V8 cylinder models | 12,80 | 22.50 pints |
| Cooling system, Turbo charged diesel models | 11,1 | 20.0 pints |

**Recommended lubricants and fluids**
**Service instructions for temperate climates – ambient temperature range −10°C to 35°C**

## Turbo-charged diesel engine

### Engine oil

Super high performance diesel oils are recommended to provide adequate protection under all operating conditions. The engine oil and filter must be changed every 10,000 km (6000 miles) and it is important that only oils specified in the lubrication chart are used. If one of the specified 'emergency' oils is used, it MUST NOT EXCEED 5000 km (3000 miles) before it is changed.

Under severe operating conditions e.g. off road in mud, airborne sand, dust, operating at high speeds in high ambient temperatures above 40°C or continual stop/start operation, the oil and filter change period should not exceed 5000 km (3000 miles). Continuous off road operation in mud, dust and wading conditions requires a monthly oil and filter change. Failure to adhere to the recommended service and operating instructions may result in premature engine wear or damage.

### Topping-up

The high performance oils recommended in the lubrication chart should be used for topping-up between oil changes, but if these are not available, the specified emergency diesel engine lubricating oils can be used, as relatively small amounts will not cause adverse affects.

### Recommended engine oils

### Recommended service instructions for temperate climates – ambient temperature range −10°C to 35°C

The following list of recommended engine oils should be used for oil changes and topping-up. These SHPD (super high performance diesel) oils allow a maximum of 10,000 km (6000 miles) between oil and filter changes.

| | |
|---|---|
| BP | Vanellus C3 Extra 15/40 |
| CASTROL | Deusol Turbomax 15/40 |
| MOBIL | Delvac 1400 Super 15/40 |
| SHELL | Myrina 15/40 |

The following list of oils is for emergency use only if the above oils are not available. They can be used for topping-up without detriment but if used for engine oil changing, they are limited to a maximum of 5000 km (3000 miles) between oil and filter changes.

| | |
|---|---|
| BP | Vanellus C3 Multigrade 15/40 |
| CASTROL | Deusol RX Super 15/40 |
| DUCKHAMS | Hypergrade 15/50 |
| ESSO | Essolube XD-3 15/40 |
| MOBIL | Delvac Super 15/40 |
| PETROFINA | Fina Dilano HPD 15/40 |
| SHELL | Rimula X 15/40 |
| TEXACO | URSA Super Plus 15/40 |

Use only oils to MIL-L-2104C/D or API Service levels CD or SE/CD-15W/40.

**Recommended lubricants and fluids**
**Service instructions for temperate climates – ambient temperature range − 10°C to 35°C**

| COMPONENTS | BP | CASTROL | DUCKHAMS | ESSO | MOBIL | PETROFINA | SHELL | TEXACO |
|---|---|---|---|---|---|---|---|---|
| Engine V8 Carburetter Dashpots | BP Visco 2000 15W/40 or BP Visco Nova 10W/30 | Castrol GTX 15W/50 or Castrolite 10W/40 | Duckhams 15W/50 Hypergrade Motor Oil | Esso Superlube 15W/40 | Mobil Super 15W/40 or 10W/40 | Fina Supergrade Motor Oil 15W/40 or 10W/40 | Shell Super Motor Oil 15W/40 or 10W/40 | Havoline Motor Oil 15W/40 |
| Engine 4-cyl. petrol | | | | | | | | |
| Engine 4-cyl. diesel – Except Turbo-charged engine | BP Visco 2000 15W/40 or BP Visco Nova 10W/30 or BP Vanellus C3 Multigrade 15W/40 | Castrol GTX 15W/50 or Castrol Deusol RX Super 15W/40 or Castrolite 10W/40 | Duckhams 15W/50 Hypergrade Motor Oil or Duckhams Fleetol Multi-V 20W/50 or Duckhams Fleet Master 15W/40 | Esso Superlube 15W/40 or Essolube XD-3 15W/40 | Mobil Super 15W/40 or Mobil Delvac Super 15W/40 or Mobil Super 10W/40 | Fina Supergrade Motor Oil 15W/40 or 10W/40 | Shell Super Motor Oil 15W/40 or 10W/40 or Shell Rimula X 15W/40 | Havoline Motor Oil 15W/40 or URSA Super Plus 15W/40 or Eurotex Motor Oil 15W/50 |
| LT77 — five-speed gearbox – 4-cylinder | BP Autran G | Castrol TQF | Duckhams Q-Matic | Esso ATF Type G | Mobil ATF 210 D | Fina Purfimatic 33G | Shell Donax TF | Texamatic Type G |
| LT85 — five-speed gearbox – V8 cylinder | BP Visco 2000 15W/40 or BP Visco Nova 10W/30 | Castrol GTX 15W/50 or Castrolite 10W/40 | Duckhams Hypergrade 15W/50 | Esso Super Lube 15W/40 | Mobil Super 15W/40 or 10W/40 | Fina Supergrade Motor Oil 15W/40 or 10W/40 | Shell Super Multi-grade 15W/40 or 10W/30 | Havoline Motor Oil 15W/40 or 10W/40 |

**Recommended lubricants and fluids (continued)**

| COMPONENTS | BP | CASTROL | DUCKHAMS | ESSO | MOBIL | PETROFINA | SHELL | TEXACO |
|---|---|---|---|---|---|---|---|---|
| Transfer box<br>Final drive units<br>Swivel pin<br>Housings<br>Steering box | BP<br>Gear Oil<br>SAE 90EP | Castrol<br>Hypoy<br>SAE 90EP | Duckhams<br>Hypoid 90 | Esso<br>Gear Oil<br>GX 85W/90 | Mobil<br>Mobilube<br>HD 90 | Fina<br>Pontonic<br>MP<br>SAE 80W/90 | Shell<br>Spirax<br>90EP | Texaco<br>Multigear<br>Lubricant<br>SAE 85W/90 |
| Prop. shaft<br>Front and rear | BP<br>Energrease<br>L2 | Castrol<br>LM Grease | Duckhams<br>LB 10 | Esso<br>Multi-<br>purpose<br>Grease H | Mobil-<br>grease MP | Fina<br>Marson<br>HTL 2 | Shell<br>Retinax A | Marfak<br>All purpose<br>Grease |
| Steering box | BP Gear Oil<br>SAE 90 EP | Castrol<br>Hypoy<br>SAE 90 EP | Duckhams<br>Hypoid 90 | Esso<br>Gear Oil<br>GX<br>85W/90 | Mobil<br>Mobilube<br>HD 90 | Fina<br>Pontonic MP<br>SAE 90 | Shell<br>Spirax<br>90 EP | Texaco<br>Multigear<br>Lubricant<br>EP 85W/90 |
| Power steering<br>fluid reservoir<br>as applicable | BP<br>Autran<br>DX2D | Castrol<br>TQ<br>Dexron IID | Duckhams<br>Fleetmatic CD<br>or Duckhams<br>D-Matic | Esso<br>ATF<br>Dexron IID | Mobil<br>ATF 220 D | Fina<br>Dexron IID | Shell<br>ATF<br>Dexron IID | Texamatic<br>Fluid 9226 |
| Brake and clutch<br>reservoirs | Universal Brake Fluid or other brake fluids having a minimum boiling point of 260°C (500°F) and complying with FMVSS 116 DOT 3 ||||||||
| Lubrication nipples<br>(hubs, ball joints,<br>etc.) | BP<br>Energrease<br>L2 | Castrol<br>LM Grease | Duckhams<br>LB 10 | Esso<br>Multi-<br>purpose<br>Grease H | Mobil-<br>grease MP | Fina<br>Marson<br>HTL 2 | Shell<br>Retinax A | Marfak<br>All purpose<br>Grease |
| Ball joint assembly<br>Top link | Dextagrease Super GP ||||||||
| Cooling system<br>Anti-freeze | Universal Anti-freeze<br>See later page for instructions ||||||||

Recommended lubricants and fluids    Service instructions all markets

| COMPONENTS | BP | CASTROL | DUCKHAMS | ESSO | MOBIL | PETROFINA | SHELL | TEXACO | SPEC. REF. ALL BRANDS |
|---|---|---|---|---|---|---|---|---|
| Windscreen hinges Ventilator hinges Ventilator control Seal slides, Hood retention clips. Door lock striker | BP Energrease L2 | Castrol LM Grease | Duckhams LB 10 | Esso Multi-purpose Grease H | Mobil Mobil-grease MP | Fina Marson HTL2 | Shell Retinax A | Marfak All purpose Grease | NLGI-2 Multi-purpose Lithium-based Grease |
| Windscreen washers | All Seasons Screen Washer Fluid | | | | | | | | |
| Bonnet pintle | Graphite Lock Grease Type 'B' | | | | | | | | |
| Door locks (anti-burst) Inertia reels | **DO NOT LUBRICATE.** These components are 'life' lubricated at the manufacturing stage. | | | | | | | | |
| | **NOTE:** The above lubricants are considered to be suitable for ambient temperatures in the range of: −40°C to +35°C  For extreme ambient temperatures, outside the above range, refer to local Distributor. | | | | | | | | |
| Battery lugs Earthing surfaces Where paint has been removed | Petroleum jelly **NOTE:** Do not use Silicone Grease. | | | | | | | | |
| AIR CONDITION-ING SYSTEM Refrigerant | METHYLCHLORIDE REFRIGERANTS MUST NOT BE USED Use only with refrigerant 12. This includes 'Freon 12' and 'Arcton 12' | | | | | | | | |
| Compressor Oil | Shell Clavus 68 | | BP Energol LPT 68 | | Sunisco 4GS | | Texaco Capella E Wax Free 68 | | |

**Recommended lubricants and fluids**
**Service instructions for ambient conditions outside temperatue climate limits**
**or for markets where the products listed are not available**

| COMPONENTS | SERVICE CLASSIFICATION WORLDWIDE | | | AMBIENT TEMPERATURE °C | | | | | | |
|---|---|---|---|---|---|---|---|---|---|---|
| | PERFORMANCE LEVEL | | SAE VISCOSITY | −30° | −20° | −10° | 0° | 10° | 20° | 30° |
| Engine<br><br>Carburetter dashpots<br>Oil can<br>Oils | Petrol<br><br>Oils must meet BL Cars spec. BLS.22.OL.02 or API service levels SE or SF or SE/CC or SF/CC or SE/CD or SF/CD or the CCMC requirements | Diesel<br><br>Oils must meet BL Cars spec. BLS.22.OL.06 and MIL-L-46152 or API service levels CC or CD or SE/CC or SE/CD or SF/CC or SF/CD or the CCMC D1 requirements | 5W/20    5W/30<br>5W/40<br><br>10W/30<br><br>10W/40    10W/50<br><br>15W/40    15W/50<br><br>20W/40    20W/50 | | | | | | | |
| Power steering reservoir | ATF Dexron IID | | | | | | | | | |
| Front and rear Axle differential Swivel pin housings LT230 transfer box Steering box | API GL4 or MIL-L-2105 | | 90 EP<br><br>80W EP | | | | | | | |
| LT77 gearbox – 4 cyl. | ATF M2C 33F or G | | | | | | | | | |
| LT85 gearbox – V8 cylinder | Oils must meet BL Cars spec. BLS.22.OL.02 or API service levels SE or SF or SE/CC or SE/CD or SF/CC SF/CD or the CCMC requirements | | 10W/30    10W/40<br>10W/50<br><br><br>10W/40    10W/50<br>15W/40    15W/50 }<br>20W/40    20W/50 } | | | | | | | |
| Brake and clutch reservoirs | Brake fluid must have a minimum boiling point of 260°C (500°F) and comply with FMVSS 116 DOT 3 | | | | | | | | | |
| Lubrication nipples (hubs, ball joints, etc.) | NLGI-2 multipurpose lithium based grease | | | | | | | | | |

**Service instructions for ambient conditions outside temperature climate limits
or for markets where the products listed are not available**

Anti-freeze:   Ethylene Glycol based anti-freeze (containing no methanol) with non-phosphate corrosion inhibitors suitable for use in
(a) cast iron engines (4-cyl.)
(b) aluminium engines (V8) to ensure protection of the cooling system against frost and corrosion.

4-cyl. engine: One part anti-freeze, two parts water, i.e min 33% anti-freeze in coolant. Complete protection down to −20°C
One part anti-freeze, one part water, i.e. min. 50% anti-freeze in coolant. Complete protection below −20°C to −36°C

V8 engine:   One part anti-freeze, one part water, i.e. 50% anti-freeze in coolant. Complete protection down to −36°C
**IMPORTANT:   Coolant solution must not fall below proportions of one part anti-freeze to three parts water, i.e. min. 25%
anti-freeze in coolant, otherwise damage to engine is liable to occur.**
When anti-freeze is not required the cooling system must be flushed out with clean water and refilled with a solution of one part Marstons SQ 36 inhibitor to nine parts water, i.e. 10% inhibitor in coolant.

**Passenger Car Fuel Consumption Order 1983 No. 1486**
**80/12 68 EEC**

| | Simulated Urban Cycle (mpg) | Constant Speed 56mph (mpg) | Constant Speed 75mph (mph) | Simulated Urban Cycle L/100km | Constant Speed 90Kph L/100km | Constant Speed 120Kph L/100km |
|---|---|---|---|---|---|---|
| Ninety 2.5 Petrol | 16.3 | 22.8 | N/A | 17.3 | 12.4 | N/A |
| Ninety 2.5 Diesel | 26.6 | 28.2 | N/A | 10.6 | 10.0 | N/A |
| Ninety 2.5 Turbo Diesel | 23.7 | 26.7 | N/A | 11.9 | 10.6 | N/A |
| Ninety 3.5 Petrol | 14.1 | 22.2 | 14.9 | 20.0 | 12.7 | 19.0 |
| One Ten 2.5 Petrol | 14.5 | 21.0 | N/A | 19.4 | 13.5 | N/A |
| One Ten 2.5 Diesel | 21.6 | 24.7 | N/A | 13.1 | 11.4 | N/A |
| One Ten 2.5 Turbo Diesel | 23.1 | 25.7 | N/A | 12.2 | 11.0 | N/A |
| One Ten 3.5 Petrol | 13.0 | 21.0 | 14.8 | 21.7 | 13.4 | 19.1 |

**NOTE**
The results given above do not express or imply any guarantee of the fuel consumption of the particular car with which this information is supplied. The car itself has not been tested and there are inevitably differences between individual cars of the same model. In addition, this vehicle may incorporate particular modifications. Furthermore, the driver's style and road and traffic conditions, as well as the extent to which the car has been driven and the standard of maintenance, will all affect its fuel consumption. Information as to the results of officially approved tests on all cars tested is available for inspection by customers on the premises where these cars are displayed. The fuel consumption figures specified in this Drivers Handbook do not allow for towing, off road use or any other adverse driving conditions.

**Electrical system**

| | |
|---|---|
| Type......................................................................... | Negative earth |
| Voltage .................................................................... | 12 |
| Battery — Petrol models ............................................ | Lucas-standard BBMS No. 371 } 9 plate |
| | Chloride-standard BBMS No. 291 } Designation 190/84/90 |
| — Diesel models.................................... | BBMS No. 372 14 plate Designation 380/120/90 |
| Charging circuit — 4-cylinder models ........................... | Alternator 115/34 |
| — V8 cylinder models........................... | Alternator 115/45 |
| Ignition system — Petrol models ................................. | Coil |

**Replacement bulbs and units**

Headlamps

| | | |
|---|---|---|
| — UK ...................................................... | 75/50 W Sealed beam unit | |
| * — Europe (except France)........................................ | 60/55 W Halogen bulb | Local legislative requirements |
| — France and Algeria........................................... | 60/55 W Halogen bulb, yellow | may require fitment of |
| — Rest of world, right-hand steering ......................... | 75/50 W Sealed beam unit } | quartz-halogen headlamps in |
| — Rest of world, left-hand steering .......................... | 60/50 W Sealed beam unit } | countries outside Europe. Refer |
| Front side lamps................................................... | 12 V 5 W | to Distributor or Dealer for |
| Side repeater lamps .............................................. | 12 V 4 W | details. |
| Stop/tail lamps .................................................... | 12 V 21/5 W | |
| Flasher lamps...................................................... | 12 V 21 W | |
| Number plate lamp............................................... | 12 V 4 W | |
| Reverse lamp ...................................................... | 12 V 21 W | |
| Rear fog guard lamp bulb ....................................... | 12 V 21 W | |
| Interior lamp ...................................................... | 12 V 21 W | |
| Warning lights (except diesel cold start) ......................... | 12 V 1.2 W | |
| — diesel cold start.................................... | 6 V 1.2 W | |
| Instrument illumination.......................................... | 12 V 3 W | |
| Hazard switch warning light ................................... | 12 V 0.6 W | |

*The 60/55 W Halogen bulb is fitted to the Land Rover 'County' Station Wagon

See fold-out
diagram at back of
handbook

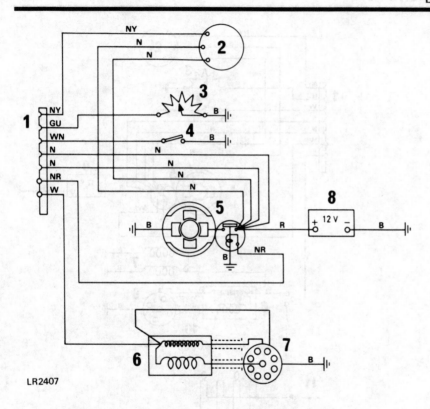

**V8 Engine – Fig. LR2407**
1. Connector
2. Alternator (A127/45A)
3. Water temperature transmitter
4. Oil pressure switch
5. Starter motor solenoid
6. Coil
7. Distributor
8. Vehicle battery

LR2407

**4-cylinder petrol or diesel engine circuit –
Fig. ST264**

1. Connector
2. Alternator (A127/45A)
3. Water temperature transmitter
4. Oil pressure switch
5. Starter motor solenoid
6. Coil (petrol engine)
7. Carburetter solenoids
8. Distributor
9. Vehicle battery
10. Diesel shut-off solenoid
11. Glow plugs

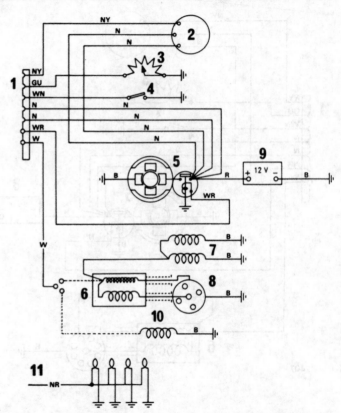

ST264

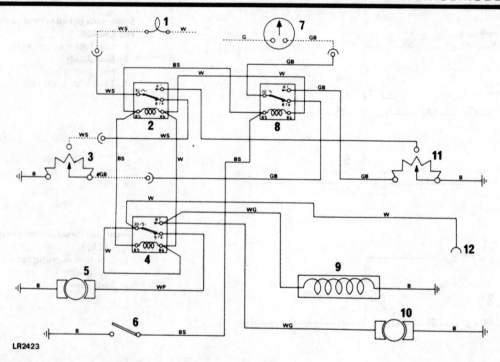

**Twin tank petrol system circuit – Fig. LR2423**

1. Low fuel warning light
2. Low fuel warning relay
3. Rear tank transmitter unit
4. Fuel pumps relay
5. Rear tank fuel pump
6. Fuel changeover switch
7. Fuel gauge
8. Fuel gauge relay
9. Solenoid valve
10. Side tank fuel pump
11. Side tank transmitter unit
12. Ignition feed main harness

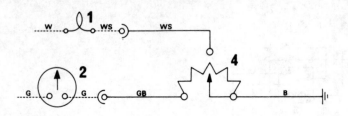

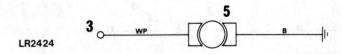

LR2424

**Single tank petrol system circuit –
Fig. LR2424**

1. Low fuel warning light
2. Fuel gauge
3. Fuel pump feed
4. Tank transmitter unit
5. Fuel pump

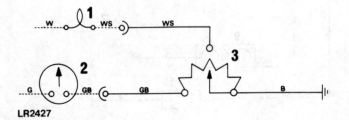

LR2427

**Single tank diesel system circuit –
Fig. LR2427**

1. Low fuel warning light
2. Fuel gauge
3. Tank unit

**Twin tank diesel system circuit –
Fig. LR2426**

1. Low fuel warning light
2. Low fuel warning relay
3. Rear tank transmitter unit
4. Fuel changeover switch
5. Fuel gauge relay
6. Side tank transmitter unit
7. Fuel gauge
8. Ignition feed main harness

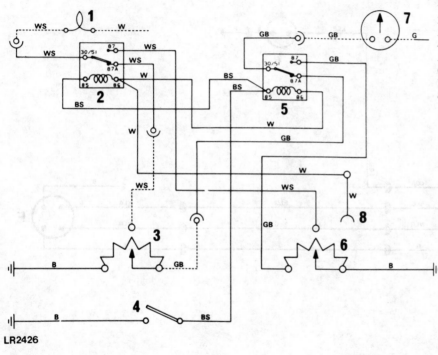

LR2426

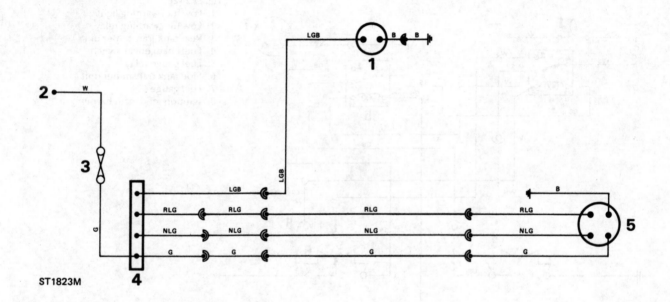

ST1823M

**Rear wash/wipe circuit – Fig. ST1823M**
1. Washer pump
2. Ignition feed
3. 5 amp fuse
4. Wiper switch
5. Wiper motor

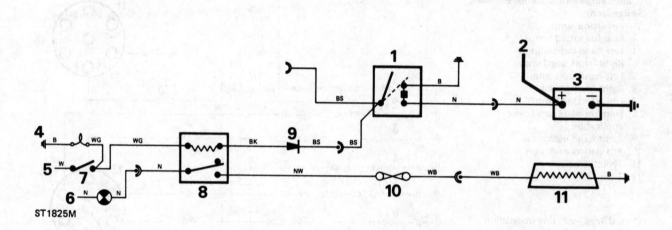

ST1825M

**Heated rear window – Fig. ST1825M**

1. Voltage sensitive switch
2. Starter solenoid
3. Battery
4. Heated rear window warning light
5. Ignition feed
6. Battery positive
7. Heated rear window switch
8. Heated rear window relay
9. Diode
10. 10 amp fuse
11. Heated rear window

**Key to trailer socket circuit – Fig. ST267
(United Kingdom and Europe excluding
Switzerland)**

1. Reversing lamp
2. Rear fog guard lamp
3. Left-hand indicators
4. Right-hand stop lamp
5. Left-hand stop lamp
6. Right-hand tail lamp
7. Left-hand tail lamp
8. Number plate lamp (where
   applicable)
9. Right-hand indicators
10. Auxiliary connection
11. Type 12S socket
12. Type 12N socket

Dotted lines – existing installation
Full lines – new leads

**Key to cable colours**

| | | |
|---|---|---|
| N Brown | B Black | P Purple |
| W White | Y Yellow | R Red |
| G Green | O Orange | |

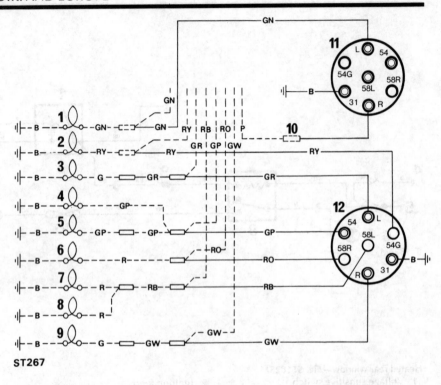

ST267

**Key to trailer socket circuit – Fig. ST268
(Switzerland only)**
1. Column switch cable assembly
2. Type DB10 relay
3. Instrument cable assembly
4. Main cable assembly
5. Cable assembly
6. Toe box connections
7. Spare position
8. Stop lamp switch
9. Right-hand rear lamp connections
10. Type 12N trailer socket

Dotted lines – existing installation
Full lines – new leads

Key to cable colours

| N Brown | B Black | P Purple |
|---------|---------|----------|
| L Light | W White | Y Yellow |
| R Red   | G Green | O Orange |

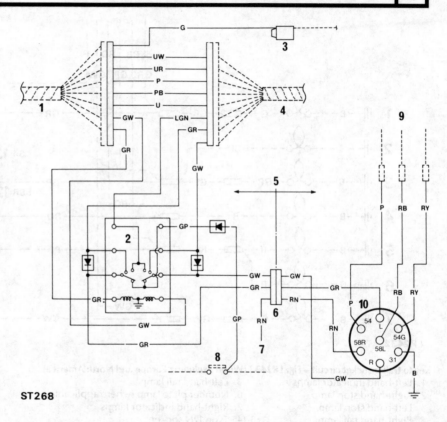

ST268

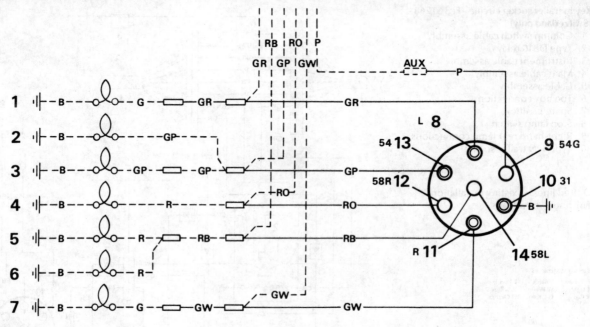

LR2432

**Key to trailer socket circuit – Fig. LR2432 (World excluding Europe and North America)**

1. Left-hand indicator lamps
2. Right-hand stop lamp
3. Left-hand stop lamp
4. Right-hand tail lamp
5. Left-hand tail lamp
6. Number plate lamp (where applicable)
7. Right-hand indicator lamps
8-14. Type 12N socket

Dotted lines – existing installation
Full lines – new leads

Key to cable colours

| | | |
|---|---|---|
| B Black | W White | P Purple |
| R Red | G Green | O Orange |